COLOGY
ND
CONOMICS
ONTROLLING POLLUTION
THE 70ˢ

Written
and
edited
by
Marshall I.
Goldman

ECOLOGY AND ECONOMICS: CONTROLLING POLLUTION IN THE 70s

ECOLOGY AND ECONOMICS: CONTROLLING POLLUTION IN THE 70s

Written and Edited by
MARSHALL I. GOLDMAN

PRENTICE-HALL, INC.
Englewood Cliffs, New Jersey

ECOLOGY AND ECONOMICS:
CONTROLLING POLLUTION IN THE 70s
Written and Edited by Marshall I. Goldman

A revised and enlarged edition of
CONTROLLING POLLUTION: The Economics of a Cleaner America

Otto Eckstein, Series Editor

DUHS
G

ISBN: P 0-13-222729-0

ISBN: C 0-13-222737-1

48

Library of Congress Catalog Card No. 75-171963

10 9 8 7 6 5 4 3 2 1

PRENTICE-HALL INTERNATIONAL, INC., *London*
PRENTICE-HALL OF AUSTRALIA, PTY. LTD., *Sydney*
PRENTICE-HALL OF CANADA, LTD., *Toronto*
PRENTICE-HALL OF INDIA PRIVATE LIMITED, *New Delhi*
PRENTICE-HALL OF JAPAN, INC., *Tokyo*

PRINTED IN THE UNITED STATES OF AMERICA

72' 00065

Contents

Preface

Since the preparation of *Controlling Pollution: The Economics of a Cleaner America* the subject of environmental disruption has become one of the major issues of the age. Much to our shame, it will probably remain an issue for the foreseeable future. This interest and urgency has made the enlargement of the book both easier and harder. Compared to five years ago, much more information has been published about the subject. While this new information increases our understanding, it also makes it difficult to hold the material that should be discussed to a manageable size.

Of all the projects that I have embarked upon in my academic career, the study of environmental disruption has been perhaps the most rewarding. The sincere concern of people in this country, the USSR, Japan, and England whom I have spoken with revives my spirits in what otherwise would be a depressing age. If man can indeed respond when he is made to see how his irrational abuse of nature can bring destruction upon the world, perhaps he may also respond someday to the irrational abuse of man by man.

ACKNOWLEDGEMENTS

Because a subject like pollution necessitates the knowledge of many disciplines, I have had to seek help from several of my associates. Above all, thanks must be expressed to Robert Shoop, Owen Stratton, Gerald Parker, and Lydia Goodhue. Otto Eckstein has also been gracious with his advice. In addition, I am grateful for the efforts of Merry Levering who worked with me on the preparation of the final manuscript, and to the students in my first class in the Economics of Pollution who taught me as much as I hope I taught them.

POLLUTION

Tom Lehrer

If you visit American city,
You will find it very pretty.
Just two things of which you must beware:
Don't drink the water and don't breathe the air.
Pollution, pollution,
They got smog and sewage and mud,
Turn on your tap and get hot and cold running crud.

See the halibuts and the sturgeons
Being wiped out by detergents.

Fish got to swim and birds got to fly
But they don't last long if they try.

Pollution, pollution,
You can use the latest toothpaste,
And then rinse your mouth with industrial waste.

Just go out for a breath of air,
And you'll be ready for Medicare.
The city streets are really quite a thrill,
If the hoods don't get you, the monoxide will.

Pollution, pollution,
Wear a gas mask and a veil.
Then you can breathe, long as you don't inhale.

Lots of things there that you can drink,
But stay away from the kitchen sink.
Throw out your breakfast garbage, and I've got a hunch
That the folks downstream will drink it for lunch.

So go to the city, see the crazy people there.
Like lambs to the slaughter
They're drinking the water
And breathing the air.

From "That Was the Year that Was." Music and lyrics by Tom Lehrer, copyright © 1965 by Tom Lehrer. Reprinted by permission.

ECOLOGY AND ECONOMICS:
CONTROLLING POLLUTION
IN THE 70s

INTRODUCTION

Pollution:
the mess around us

MARSHALL I. GOLDMAN

INTRODUCTION

Today's news media devote almost as much attention to air and water
pollution as to the problems of poverty. Virtually overnight pollution
seems to have become one of America's major issues. To the economist
the problem provides a unique opportunity to see the result of the diver-
gence of social and private benefits from social and private costs or what
economists call external diseconomies. The destruction of natural resources occurs largely because of the difficulty of imposing direct costs
or monetary responsibility on the polluters. Attempts to prevent pollution
illustrate how economics along with politics and science can be utilized
to cope with an increasingly dangerous situation.

Until recently few individuals or organizations were concerned with
the problems of pollution. The only previously interested ones were a few
conservationists and hikers and mountain climbers. Why, then, is there
this sudden clamor about pollution?

Pollution Through the Years

Pollution has plagued mankind for centuries, and we all know what
it is. But the word itself is often used to cover a variety of sins. Before
we can talk intelligently about it, we should try to define what we mean.
Actually, what we have had is environmental disruption. There is virtually no naturally pure water and air and most of us would not know
what to do with pure air and water if we had it. Pure oxygen, if that is

Portions of this section appeared in the article "The Costs of Fighting Pollution" and are reprinted with permission from *Current History*, August, 1970.

what we mean by pure air, would make us giddy, and pure distilled water would have no taste.

Even if pure air and water had existed at one time in a natural state, the mere presence of human beings and animals would be enough to alter those conditions. Such changes, however, are not serious unless they radically disrupt the existing balance or make the environment unfit for other organisms. Even then, it is necessary to distinguish between *impurities*, which make water and air economically or aesthetically undesirable—and may also destroy some forms of flora or fauna—and *contaminating substances*, which endanger the health of human beings.

That undesired changes were taking place in the water supply was recognized by the Romans before the first century B.C. Because the sewage generated by a city of about one million people endangered the drinking water, the Romans built one of the first major municipal sewers in history, the Cloaca Maxima. Venice, after all, was and is nothing but a sewer in search of a city. Until new construction recently disrupted the age-old water circulation pattern, the city flushed its sewage effortlessly into the sea twice a day by using the natural flow of the tides. In the north, Richard II of England as early as 1388 banned the throwing of filth into the Thames. Later generations, however, were ignorant of such necessities and dumped their sewage whenever and wherever convenient. The result was dysentery and periodic epidemics of such diseases as cholera and typhoid.

Smog has also annoyed man for centuries. Although smog in Los Angeles did not become a burning issue until 1943, in the mid-sixteenth century Spanish explorers landing there noted layers of smoke from Indian fires hanging above the area. In the scientific terminology of today this is called an inversion layer. For centuries England has been similarly plagued by polluted air. As early as the thirteenth century English authorities complained about smoke from coal and charcoal fires. By the fourteenth century the first clean air legislation had been passed, and one man was actually hanged in London for violating the law.[1] Apparently such laws later fell into disuse. This is indicated in an appeal addressed to Charles II in 1661 by John Evelyn, entitled *"FUMIFUGIUM or the Inconvenience of the Aer and SMOAKE OF LONDON. Dissipated Together with some REMEDIES humbly proposed By John Evelyn Esq: To His Sacred MAJESTIE and To the PARLIAMENT now Assembled. Published by His Majesty's Command."* Charles Dickens' England continued to suffer from sunless skies and smoky moors. Ironically, the Eng-

---

[1]Secretary of State for Local Government and Regional Planning, *Protection of the Environment* (London: Her Majesty's Stationery Office, May 1970), Cmnd 4373, p. 6.

lish government during World War II encouraged smoke emission in order to obscure bombing targets from the Germans.

Still, despite isolated examples, until the beginning of the twentieth century man has been able to coexist with his waste. There was little or no interference with nature's self-regenerating system. Generally harmonious proportions were kept among such living organisms as human beings, vegetation, and animals. Moreover, since there had been little experimentation with the earth's minerals, almost all waste decomposed rapidly or served as nutrient or raw material for other forms of life. For example, kitchen garbage was nicely disposed of by the livestock usually kept for human consumption. The circle seemed to be perfect.

By the late nineteenth century, however, the careful observer could find some disproportion in the circle. Pollution, which for centuries had not been especially offensive, gradually became intolerable in a growing number of places. In the years following World War II there was no longer any doubt that the circle had popped and left gaping holes out of which an increasingly alarmed population gasped for fresh air and water.

An Annoyance Becomes a Crisis

In this country the emergence of concern about environmental disruption has been caused by a combination of developments such as (1) population explosion, (2) unparalleled affluence, (3) technological progress, and (4) major incidents affecting the health and well-being of large numbers of people.

1. Since World War II the world's population has been expanding at an exceptionally rapid rate. Paul Ehrlich of Stanford University points out that world population doubled during the two hundred years between 1650 and 1850. The next doubling took eighty years, and now it takes only thirty-five years. Moreover, it is not just that there are more of us, but that we are clustered in ever more compact areas. Kingsley Davis estimates that 40 percent of the world's population live in urban areas; 50 percent of urban dwellers live in cities of 100,000 or more. Naturally the more people there are, the more wastes there are to recycle. The concentration of people and their accompanying wastes makes the task of recycling and disposal all the more difficult. In the past, water, air, and solid wastes were discharged in small, easily diluted doses; now wastes are collected in large sewage complexes, tall chimney stacks, or sprawling dumps that are invariably overtaxed. The same disposal problem was created when livestock was taken from wide-open ranges and farmlands and herded into feed lots. Although once manure served as the main source of fertilizer, now much of it has become a mess to be disposed of

as expeditiously and odorlessly as possible. Chemical fertilizer now accomplishes artificially what used to be done naturally.

To escape the crushing throngs in the cities, people formed a mass exodus to the suburbs. Large numbers of people apparently decided that they did not like being crowded together in the city; if they were going to be crowded, they decided that it was much better to crowd together in the suburbs. Green spaces began to disappear; before long airplane pilots and geographers found themselves unable to distinguish where one town ended and another began. Vast portions of the countryside from New Hampshire to Virginia became one continuous suburb, which Jean Gottman called "Megalopolis." This term may also come to describe the area from Los Angeles to San Francisco and from Chicago to Detroit or from Chicago to Pittsburgh.

As the population grew and increased its mobility and its discharge of refuse, natural facilities for transforming waste began to disappear. Factories and stores moved outside of the cities, brought with them highways, asphalt parking lots, and demands for water and air. To the regular wastes of consumption and production in the cities were added the remains of buildings demolished under urban renewal programs. In many areas it became harder and harder to find natural preserves in which to process or absorb smoke and liquid and solid wastes.

2. The population has grown not only in size but also in wealth. With per capita income reaching historic heights, production has risen to satisfy growing demands. This production increases consumption of natural resources for industrial purposes, which in turn generates greater waste from the process of manufacturing and mining itself.

It is not only from the production of goods that wastes are multiplied; it is also in their consumption. The richer we become and the more we can consume, the more we have to throw away. Furthermore we somehow decided that it is now convenient to buy something and throw either it or its container away. Today almost everything is disposable, from diapers to dresses. We now even have disposable disposables. Unfortunately the use of such disposables is not the final solution to our solid waste disposal problem. Our cast off purchases must be put somewhere. Even if a product is utilized for a long period of time (a piece of furniture, a car, a television set), ultimately it will be discarded in some form—usually as junk. Ultimate disposal results in the disintegration of a product into its elemental components, as with the digestion of food or the burning of fuel. Our wastes must end up in the water, on the ground, or in the air, generally in a form making them unsuitable for further use (see Part II). With our gargantuan appetites, Paul Ehrlich argues, each American has about fifty times the negative impact on the earth's life support system as does the average citizen of India. No wonder some have come to call America the "effluent society."

The magnitude and complexity of the disposal problem is best illustrated by the complications that have arisen out of the production and consumption of the automobile. Affluence, the automobile, and aggravation seem to go together. For example, automobile exhaust is now the major cause of air pollution in many of our larger cities. Annual automobile production has risen from 3.5 million in 1947 to 9 million in the 1960s. The fumes produced annually by the almost 110 million combustion engines of America's automobiles, trucks, and buses are estimated to be 90 million tons of gases, including 64 million tons of carbon monoxide.[2] This constitutes 40 percent of all air pollution emissions and 60 percent of all carbon monoxide released into the atmosphere. The carbon monoxide alone is enough to poison the combined air space of Massachusetts, Connecticut, and New Jersey. The automobile is the king of American consumer goods, and it is also a complete portable factory. Every vehicle generates power and air exhaust just like a miniature thermal electric plant. Crossing behind a jam of cars stopped at a red light is a bit like walking behind a series of upended chimney stacks—all pointing in your direction.

Some cities have been particularly hard hit by automobile pollution. Because of Los Angeles' topography even the slightest amount of gas or smoke has a severe effect on the purity of the air. The prevailing winds are not very strong, and those that come from the ocean are unable to carry the smoke over the hills surrounding the city. As a result Los Angeles has passed exceptionally rigid laws that have limited and almost stopped smoke from factories and municipal incinerators. Unfortunately, Los Angeles has had much less success in controlling automobile exhaust. Vehicular exhaust is now responsible for approximately 80 percent of Los Angeles' smog. For many years it was felt that smog caused by automobiles was unique to Los Angeles because of its geography and large automobile population. Recently the problem has spread to almost all large metropolitan areas. No longer is Los Angeles the only city in the country where you wake up in the morning and hear the birds coughing!

Of equal importance, automobile junking, like the disposal of other consumer goods, has become a serious problem. Every year approximately 7 million vehicles are scrapped and must be discarded somehow. Fortunately many cars and trucks are recycled in the form of used cars and trucks a few years after their initial purchase, but ultimately even these vehicles must be removed from the streets. At one time this re-

[2]Council on Environmental Quality (CEQ), *Environmental Quality,* First Annual Report (Washington, D.C.: Government Printing Office, August 1970), p. 63.

moval posed no difficulty; but beginning in the late 1950s, automobiles started to pile up in city streets and in automobile graveyards across the country (see Table I). The metamorphosis of the discarded auto from a depreciated but still treasured pet into a valueless white elephant is a classic illustration of how changes in technology and economic factor costs combine to effect environmental disruption.

Until the early 1960s steel was produced almost entirely in open-hearth furnaces. The metal charge fed into these furnaces often consisted of as much as 50 percent scrap iron along with the raw iron ore. Accordingly, used automobiles were sought after by junk dealers because of the value of the scrap metal. Despite the fact that other metals, such as copper and aluminum, were sometimes included along with the scrap steel in the melted mass, the technology of the open-hearth furnace was such that the quality of the steel it produced was not seriously affected.

Because both its construction and operating costs were often one-half as large as the oxygen (LD) process of making steel, more and more American steel manufacturers began to close down their open-hearth furnaces and replace them with oxygen furnaces. Unhappily for the junk dealer, oxygen furnaces had a much lower tolerance for impurities and could take no more than 30 percent scrap in their metal charges. Inevitably the steel manufacturers began to substitute iron ore and taconite pellets for scrap. This in turn caused a drop in the price of scrap and the value of the junked auto.

For a time it was still worthwhile for junk dealers to process old cars and recycle the ferrous metals. If they were willing to sort out the nonferrous metals and other nonmetallic scrap, the reclaimed steel was still valuable. The most common way to do this was to burn off as much waste as possible and to sort by hand those parts that remained. As long as smoke control laws were lax and labor costs were low, the junk dealer could make a profit on such operations. However, new prohibitions on the open burning of junked automobiles and rising labor costs sharply

Table I: *Abandoned Cars Towed Away in City of New York*

Year	Number of Cars	Year	Number of Cars
1960	2,500	1966	23,795
1961	5,117	1967	25,842
1962	6,299	1968	31,578
1963	13,579	1969	57,742
1964	23,386	1970	72,961
1965	21,943		

Source: *New York Times,* January 24, 1971, p. 3.

increased the junkman's costs. When combined with the fall in the price of scrap from about $40 a ton to $25 a ton, the increase in costs made it unprofitable for the junkman to buy up junked autos. Some dealers even began demanding fees for hauling away scrapped cars. Predictably, this led to the widespread practice of abandoning cars in countless unexpected places. In New York City alone, over 72,000 cars were abandoned on the city's streets in 1970 (see Table I); in Chicago there were 50,000. Now it became the cities' task to haul the cars away to automobile graveyards or to pay a bounty to the junkman to do so.

The increase in national wealth produced a keener awareness of pollution for yet another reason. More wealth brought with it not only more products but also more leisure, making it possible for people to become aware of and to explore the countryside. More and more they found that what was left was becoming polluted. Invariably, this made an indelible imprint. A famous Michigan labor leader unexpectedly asked to testify before antipollution hearings conducted in Detroit. Since this particular leader had never taken much interest in pollution control before, he was asked to account for his sudden enthusiasm. He sadly explained that after years of hard work, he finally managed to buy a summer retreat. To his dismay, within a few years, the lake adjacent to his cottage had become polluted; from Shangri-la to cesspool. As one government official explained it, "This is how we win our most ardent supporters."

3. A third factor influencing the seriousness of the pollution problem has been the rapid advance in industrial technology. It is not just that there are more of us and that each of us consumes more than did our fathers and grandfathers; it is also that what we consume is more complex in its material makeup. Each day, products of an ever more exotic and synthetic nature are invented and distributed. It is estimated that 500 new chemical compounds are introduced by industry each year. Frequently these newly-discovered compounds are not biodegradable—readily broken up into easily digestible or disposable by-products. We may live better thanks to chemistry, but the products that result often live on long after the users are gone. Some products, such as aluminum tin cans, are virtually indestructible. The old steel tin cans at least rusted and disintegrated after a time. Unhappily, manufacturers are moving further and further away from products and containers like ice cream and its cone, ideal from the point of view of pollution control because they self-destruct in the process of consumption. Instead, both manufacturers and consumers are encouraging the use of permanent plastic cups and nonreturnable bottles, which means greater convenience for the consumer but more litter on the picnic ground and roadside.

The tendency is to blame manufacturers for the switch to nonreturnable bottles. In fairness, it is as much if not more the fault of the consumer. The bottlers began to switch when they found that their customers were failing to claim their deposits for returnable bottles. From an average of fifty round trips per bottle in 1955, returnable bottles were making only about ten trips in 1970.[3] Many bottlers simply abandoned the use of returnable bottles altogether. Thus there is no longer monetary compensation for most bottle returns. The withdrawal of the economic incentive, small as it is, has simply accelerated the accumulation of clutter.

The effects of technology have been even more serious in other fields. Deadly pesticides and industrial wastes are composed of chemical derivatives that do not always break down easily. That, in fact, was what once made them so attractive. DDT sprayed on a plant did not dissolve and lose its effectiveness. As a result, malaria, one of the worst scourges of man, was virtually eliminated in vast parts of the would and DDT was regarded as a major boon to mankind. Only recently have people like Rachel Carson convinced us that such compounds disrupt ecological processes and become ruthless killers of plants and wildlife. Occasionally such compounds build up rapidly in the food chain and end up in the human body; lead and mercury poisoning are two of the more familiar examples of what can happen.

In a fascinating article in the April 1971 issue of *Environment,* Barry Commoner, Michael Corr, and Paul J. Stamler argue that indeed the rapid increase in environmental disruption is due primarily to the switch in technology, which has brought about heightened reliance on synthetic chemicals, the private automobile, cement, and domestic electrical appliances, and much greater power consumption. With the aid of a rudimentary index, they try to show that environmental conditions have deteriorated much more rapidly than could be explained by increases in either population or affluence. In contrast, the increased production of synthetics like plastics and artificial fertilizer and aluminum is on a par with the increased pollutants they find in the environment.

The soap industry is a prime example of how the thoughtless use of technology can disrupt the environment. In an effort to improve the cleansing impact and lower the cost of the product, the soap industry switched from a fat base to a nonbiodegradable detergent base—one that does not readily break down in the normal course of sanitary treatment. Consequently, pools of froth began to cover drinking water reservoirs, and floods of suds frequently returned to the household through the kitchen faucet.

[3]*Wall Street Journal,* June 30, 1970, p. 25; January 7, 1971, p. 1.

We should remember that the initial stimulus to the development of both pesticides and detergents was a positive one. The well-being of mankind was to be advanced through the use of technology. It was assumed on the one hand, that health could be improved by eliminating the causes of disease and on the other hand, that costs would be reduced and convenience improved by a switch to labor-saving but capital and environmental intensive products and processes. An important lesson to be learned from the outcome is that our heightened ability to tinker with technology may sometimes lead to unexpected and unfortunate results. The likelihood of such unanticipated consequences becomes all the greater as we industrialize and our ability to manipulate technology increases. In our efforts to improve on nature we sometimes find ourselves upsetting the ecological balance with potentially disastrous consequences. Invariably the engineers rush to our rescue with new solutions to the new problems we have created, but in a short time we often find that their newest solutions have in turn generated a new set of difficulties.

A good illustration of how the best-intentioned efforts to improve on nature can adversely affect the well-being of a whole area is the construction of the Aswan Dam. Initially the dam was designed to harness the water of the Nile River for irrigation and electric power generation and to prevent the annual flooding that had gone unchecked for centuries. But the Egyptians and Russians who financed and supervised the dam's construction now find that the dam has given birth to a new set of dilemmas that may eventually outweigh the benefits provided by the dam. For example, now that the flooding has been checked, the land downstream no longer benefits from the fertilizing silt that the river once provided. As a result, artificial fertilizer must be manufactured and supplied, in turn necessitating use of a good portion of the electricity being generated by the dam. The cessation of the silt flow has also prevented vital nutrients from reaching the mouth of the Nile where an important fishing operation has been carried on for centuries. The sardine catch, which was 18,000 tons in 1965, dropped to 500 tons in 1968. Not only has the sardine fishermen's livelihood been destroyed, but the fish that feed on sardines and on which the sardines feed are also being seriously affected. Moreover the curbed flow of the Nile into the Mediterranean has caused a rise in its salinity that is already beginning to affect the whole eastern end of the sea. Salt water is also spreading into the Nile River Delta from the Mediterranean and Red Sea, destroying water supplies and making farmlands saline. At the same time, with the silt removed, the Nile flows much faster below the dam so that vast new forms of erosion have been created in and along the river bed. This erosion has necessitated plans for the construction of a new series of restraining dams to slow the water flow.

The cost of this remedial project is expected to be at least $250 million. The formation of Lake Nasser, a large body of relatively stagnant water behind the dam, has given rise to the spread of bilharzia and malaria. Bilharzia, also known as schistosomiasis, is a severe intestinal disease spread by the snails that now circulate through the new irrigation canals. At the same time there is concern that the silt being trapped behind the dam at a rate of 100 million tons a year will accumulate so rapidly that the storage capacity of the dam will be sharply reduced. There are also signs that millions of cubic meters of water a year are seeping out of the basin behind the dam with as yet unknown consequences. Because of the greater exposure of the water in Lake Nasser to the sun, evaporation has also increased rapidly. This process has been accelerated by the rapid growth of water hyacinths, which not only clog much of the lake surface but also absorb much of the lake water through their leaves. This heightened evaporation could affect the climatic patterns of the region.

To the supporters of such projects, the ecological aftereffects come as quite a surprise. The dam was originally conceived of as nothing but a big plus. Now it is unclear if the overall effect is positive or negative. The balance in the future, when still unanticipated difficulties may arise, is even more uncertain. We must recognize that as we increase our technological ability to "compensate" for nature's shortcomings, we are likely to find that the potential for negative by-products may increase even faster.

4. Finally, concern about environmental disruption has attracted additional support after a sequence of serious incidents. Because medical science has found more and more cures for our more traditional disease-causing enemies, we have become more prone to other ailments. Not surprisingly therefore, many people, especially the elderly, find themselves becoming affected more and more by pollution in the environment, especially the air. Thus poor air has been blamed for cancer, pneumonia, bronchitis, emphysema and tuberculosis. Scientists point out that by breathing the air in New York City, one inhales an amount of cancer-producing benzopyrene equivalent to smoking one or two packs of cigarettes a day. It is further claimed that air pollution is responsible for the 80 percent rise in deaths from respiratory diseases from 1930 to 1960.

Clearly delineated surges in the death rate have been traced directly to air pollution. Such incidents have occurred in Donora, Pennsylvania, in October 1948 when the death of 19 and the illness of 6,000 of the town's 13,800 citizens were blamed on smog. Smog was also blamed in London when the normal death rate rose by 4,000 in December 1952, and when 8,000 died prematurely in January and February 1953. Similarly air pollution was considered a major cause of death in New York

City in January and February 1963 when there were 647 more deaths than normal. Some scientists also cite air pollution as a factor contributing to street riots in some of our large cities. In addition to the normally expected discomfort, pollution is said to generate depression and melancholia. As for water pollution, in 1965 at Riverside, California, 18,000 people were afflicted with gastroenteritis from the town's water wells, and three apparently died as a result. In Japan forty-five residents of Minamata died from mercury poisoning of the water in the 1950s and 1960s (see Part III). In the 1970s fears of similar incidents spread throughout the rest of the world as some fish were discovered to have excessively high mercury content. Even when no loss of life results, the massive environmental disruption that stems from disasters such as the sinking of the Torrey Canyon or the oil spills off Santa Barbara focuses world attention on the destruction we can inflict on one another, and more than anything else heightens our awareness of the problem.

Pollution and Industrialization

It is worth repeating that despite the immense concern we in the United States now have, "effluence" is not just a by-product of the modern affluent society. For that matter, pollution is not a problem peculiar to only certain locations or types of societies. As is indicated in Part IV, pollution is not restricted to developed capitalist countries. The Soviet Union, like other countries, has its pollution difficulties. There is reason to believe that it is only the very wealthy countries, able to afford the luxury of clean water and air, that can make a fuss about it. It is expensive for industry, whether it be state or privately owned, to control pollution. Usually pollution prevention is regarded as a nonproductive expense not very much different from charity. Both expenditures are characteristic of civilized living, but expendable luxuries nonetheless which have the effect of reducing the manager's net profit. Despite the logic of economic theory which teaches that only privately owned factories pollute at the expense of the general public, state enterprises in Eastern Europe pollute with generally less restraint than private enterprises in the United States. In communist society, the factory manager is usually just as eager as his American counterpart to show as high a net profit as possible and so he too tends to neglect pollution control. But in the United States, the government frequently interposes itself in the problem of pollution because its interests are not always the same as that of the polluters. In the Soviet Union, the government is generally an economic partner of the state-owned factory and therefore not as much of an external agent as government is in the United States. Any expense that detracts from the performance of the Soviet factory in the region similarly detracts from the economic performance of the overall region. Therefore political and party officials frequently find themselves more in agreement with the polluters than with the conservationists.

There is good reason to believe that pollution control is more a function of income per capita than of socialist ownership of the means of production since the problem also confronts countries which are just beginning to industrialize. Valuable resources which should be used to control pollution are used more profitably in the increase of production. A study of environmental disruption in Japan illustrates just how difficult it is to divert resources from production (see Part III). Moreover, since production processes in the developing countries are usually less sophisticated, they usually spew out more waste per unit of product and are therefore more of a menace. Municipal water treatment in Africa and Asia and parts of Latin America is particularly deficient. However, in most industrially developing countries, the problem does not seem to be as serious and does not provoke an outcry. This may be due to the still limited stage of industrial development, at which stage industrial wastes are not by themselves enough to cause concern. In addition, as in Japan, no one wants to frustrate industrial production any more than is already the case. But precedents are being set which become difficult to alter. Manufacturers who begin production at a later time will expect the same kind of concessions since the diverting of resources for pollution abatement would put them at a competitive disadvantage. Consequently only when the state decides it is rich enough to afford some production sacrifice for cleaner air and water can there be any initiative for pollution control from the state or from industry itself.

WHY DO POLLUTERS POLLUTE?

It is not difficult to understand why there is pollution of the air and water and even the land. Moreover it is not just industries and municipalities that are contributors. Some of your best friends insist on using their own septic tanks or cesspools rather than linking up with a public sewer system for the disposal of household sewage. Unless the soil conditions are suitable and the homes are located at sufficient distance, this could contaminate the ground water. How many of your friends also burn trash in a home incinerator? Does anyone you know burn leaves in the backyard or in the street? When did you have your last cookout or throw a beer can on the beach or highway? Did you ever leave the car motor on while running an errand?

Few people stop to consider that they are polluters when they do such things. And if they recognize that they are creating some form of air, water, or solid litter, they always say to themselves that their little bit of waste will not make much difference. In addition it would cost more to have someone haul the trash and leaves away; it might be prohibitively expensive to link up a house with a sewer system. Similarly, industries and municipalities follow the same reasoning: it is cheaper, it is more convenient. Economists call the pollution which arises from such a situation an *external diseconomy*. To the polluter there never seems to be great harm in just his little pollution, but there may be great expense to society as a whole when it tries to clean up the wastes of many such polluters.

The waters are usually so plentiful and the sky so vast that it is hard to believe one person can cause pollution. In fact, one person usually cannot cause pollution; it is when there are numerous "one persons" who all think the same way that pollution results. Moreover when one person pollutes, he himself is not usually the one to suffer or bear the expense. In fact the private cost to the individual is most often cheaper if he does pollute than if he has to use expensive disposal equipment. If the water is polluted, it is the people downstream who are affected. If the air is polluted, it is the people in the direction of the wind. This is an instance where the private costs of pollution are almost always less than the costs to the rest of society. Pollution results from pushing off what should be your responsibility onto someone else who is usually anonymous. The effects are spread over such a wide area and affect so many people that generally no individual suffers enough damage to induce him to exert the effort needed to seek out and collect from the offender. Furthermore, the compensation that might be obtained from suing the offender and preventing future damage may not be worth the effort and cost involved.

There are exceptions. This becomes apparent if we divide the social costs into two categories: those that fall on the population as a whole and those that fall primarily on other producers. When the social cost of an operation is spread thinly over the entire area, no one may feel sufficiently damaged to take corrective action. However, if the smoke or discharged water from a particular factory moves downwind or downstream to an adjacent factory, the adjacent factory may feel the effects so severely that the polluted air or water becomes a direct cost of operation. In this case, the downstream or downwind factory is more likely to take action to force the offender to absorb and bear his own social costs.

This can lead to some peculiar situations. The Everett Station of the Boston Edison Company draws on water from the Mystic River for use in its boilers and for cooling. Unfortunately, a Monsanto chemical plant is located slightly upstream from the generating plant and discharges various chemical wastes into the Mystic River. This makes it necessary for Boston Edison to provide extra treatment for the water it uses. Therefore Boston Edison supports all efforts to force Monsanto to clean up its effluent. At the same time, however, until recently Boston Edison dragged its feet when others urged a reduction in sulfur and particulate content of the air discharged from its chimney stacks. To do this, it argued, would increase its costs of operation.

For the most part, however, it is usually quite difficult to assign and assess responsibility for damages in cases of external diseconomies. For example, if a second factory decides to locate next to a factory that has been polluting for decades in splendid isolation, away from all industry and homes, should the original factory be made to bear the cost of cleaning up the pollution? Why should this factory suddenly be made to pay for something it has been doing for years without visible or sufficient economic damage or complaint until the neighbor moves in alongside? Moreover if several polluters are located along a lake or are affected by one another's smoke when the wind shifts, who should bear the responsibility —the most recently completed factory? Its smoke or water may have caused the ecological system to break down, but is it ultimately to blame? In other words, who is responsible—all the men who have put straw on the camel, or just the last man before its back is broken? Clearly it would be unfair to compel the new firm to purchase expensive pollution control equipment and allow the older firms to produce waste as freely as before. This would put the new firm at a competitive disadvantage and would also discourage the location of new industry in the neighborhood. Yet if the old firms were required to buy pollution control devices, they might move out to other areas which were more permissive. Similarly, factory management could argue very persuasively that even if hundreds of thousands of dollars were spent to clean up wastes, the pollution problem might still exist unless the other firms and towns in the area were also forced to clean up their waste. Management would have to justify such expenditures to stockholders who might wonder if such expenses are warranted when they add absolutely nothing to sales revenues. Viewed from the perspective of the industrial unit such expenses are nonproductive and, if anything, benefit primarily the downstream or downwind company or town.[4]

Obviously in such circumstances the traditional price mechanism that we rely on to guide us in the production and distribution of most of our natural wealth can do nothing to solve the problem. Normally

---

[4]Allan V. Kneese, *The Economics of Regional Water Quality Management* (Baltimore: The Johns Hopkins Press, 1964) pp. 46, 62.

through the market system, prices are used to balance off the demand pressure for a product with all the production costs involved. In this way goods can be purchased by those who are willing and able to pay a high enough price for them. This, we say, shows that private costs equal private benefits. It should be pointed out that the concepts of private costs and private benefits are not peculiar to nonsocialist societies. As shown in Part IV, they also exist in the USSR, since Soviet enterprises and cities are held responsible *only* for their private costs, or charges that can be specifically and directly assessed to users of the input. As in the United States, no one in the USSR has been able to break down social costs and attribute them precisely to their source. And despite a faulty price system in the USSR, when two inputs are equally productive, the factory manager in both the United States and the USSR will use more of the input that is cheaper. For example, if wages are relatively cheaper than rents, he will use more labor than land. In other words, given a certain level of production, a manufacturer will use that mix of inputs that will cost him the least.

The price a factory manager is willing to pay also indicates that he usually expects to receive at least that much benefit from the product. Each productive resource is used until the cost of an additional unit is just equal to the extra revenue that resource will bring to the producer. When all producers react in the same way, then the price that must be paid for the productive resource will tend to equal the value of the product which is produced.

If demand for a good increases, the price will be increased. This in turn should induce an increase in production. Prices will rise also if the raw materials used in the production process become harder to find, or if workers suddenly decide they would rather work elsewhere. When the costs of production factor inputs are increased, the manufacturer tries to pass this on to the consumer by raising the sale price. It also happens in most instances that when the cost of production rises, it is necessary to cut back production to make sure that there is no overproduction. Higher prices usually mean fewer purchasers.

It is also essential, if the price system is to function properly, that someone or some group control and sell or otherwise allocate all the factors of production. Thus, whether they be landlords, bankers, laborers or managers, fees are collected for the use of an input which presumably provide proper compensation for the cost of reproducing the article or for the revenues foregone by not taking advantage of opportunities available elsewhere. In this way, the costs of all inputs used in the production process are properly identified. Thus we can say that private benefits and costs equal social benefits and costs since no resource is utilized unless someone is compensated for it at a rate which usually will reflect the demand and supply pressure for the good.

One of the reasons that air and water pollution is hard to control is that no person or organization is normally considered to be the owner of the air and water. With few exceptions such resources are considered to be readily available to those who want to use them. It is difficult to attach a value to either clean or dirty air and water. They are often treated as if they are free goods.

The price system cannot be expected to function properly, however, if the factor inputs are priced improperly or are not priced at all. If a factor is overpriced, this will lead to the increased price of the final product and a reduction in the amount sold. Conversely, if the factor is underpriced, then the selling price will not reflect the full cost to society of all the inputs that are involved, and more of the good will be sold than should be. At the same time, more of the resource will be used in production than if only the price of the factor properly reflected the alternative uses to which it might be put.

Abuse of water and air resources comes about then largely because air and water are undervalued. Air and water are regarded as free goods. For example, officials in New York City have long resisted the plea of economists and conservationists that meters be installed to measure all household use of water. Since there is no economic incentive to conserve water, there is great waste. In other words, no effort is made to "economize." ithout exception, the failure to attach a proper value to something like water results in increased consumption. When water meters and water charges were introduced throughout all of Philadelphia in 1960, water consumption declined by as much as 28 percent. All too frequently we have failed to distinguish between rain and readily available drinking water and between air and fresh air. As one critic has said, it is as if there had been no differentiation between grass and milk.

The mere imposition of a fee for the use of air and water, however, does not guarantee an end to their irrational use. There is still the risk of excess consumption if the fee is not directly related to the benefits or cost of using the water. This can best be illustrated by an example. Assume that 100 people agree to share a phone and the phone bill. Each person would feel strongly tempted to make an unlimited number of long-distance calls because he would have to pay only 1/100th of the bill. Unfortunately, each of the other participants would feel exactly the same, and the result would be a flood of long-distance calls and an unusually large phone bill. We abuse our air and water resources in exactly the same way. Even when there is a fee for water usage and sewage discharge, the fee is often understated or subsidized with income from some other source. If, however, the price of water were increased, it would become a more valuable commodity and therefore more care would be taken to utilize it efficiently.

Until recently, there has always been enough drinking water and fresh air available in the United States. There are usually alternative sources available even with water except perhaps in the West. Accordingly, in almost all other areas of the country, water and air have had no value which could be translated into dollars and cents. Hence in many areas of the United States there is an economic rationalization and even a stimulus for pollution. Input costs are understated and more air and water are used for production than necessary. Because water and air resources are underpriced, private costs are less than social costs. Under the circumstances, as even the conservative economist Milton Friedman points out in Part II, unless additional steps are taken, there will inevitably be a misallocation of society's resources and more pollution than society would otherwise deem desirable.

THE SOCIAL COSTS OF POLLUTION

Because they are generally considered to be nonmarketable products, it is difficult to evaluate the costs of polluting water and air. Cost in this context refers to the *damage* that arises because of environmental disruption; the question of how much it would cost to *remedy* environmental disruption will be considered next.

Cost estimates of damage caused by pollution are hard to make. Various estimates are tossed around, but generally they are not based on solid research or calculation. Moreover once a figure is suggested by someone, it is often grabbed by everyone else until gradually it is accepted as the basic truth. In time the original source and the qualifications surrounding the calculation are forgotten and only the basic figure remains. Yet before there can ever be a workable solution to the pollution problem, some estimate of the cost of the damage from pollution must be made so that proper values can be assigned to air and water. The task is made even more complex because so many things affected by pollution—swimming in a river or smelling clean air—are impossible to price. Even when pricing the damages is no impediment, a decision must nonetheless be made as to how far back the researcher should go in counting up the damages. Does one include just the primary effects or the secondary and tertiary damage as well? If smoggy air necessitates the closing of an airport, should one count not only the loss in revenue from the inability of the planes to fly but also the business transactions that were not consummated because salesmen were unable to reach waiting buyers?

It also happens frequently that damage arising out of some act only becomes apparent years later. For example, it has taken about twenty years for us to realize the extent of the harm caused by the use of DDT. We should reckon from the first use of the insecticide if a proper accounting of the costs is to be made. Also it is not always easy to ascribe correctly an economic loss to pollution. A house in a factory district may be priced low, but how much of this is due to pollution and how much of it is due to environmental factors which might also exist in the area? Finally, how much crime and other abuses are due to pollution which hastens the exit of more socially responsible elements of the population?

There are countless illustrations of the difficulty of providing a measure of the cost of pollution damage. Because of emissions of industrial waste into Lake Michigan from Chicago's South Side, it was necessary to close several beaches on the lake in 1965 and 1966. It turned out that blacks were the predominant users of two or three of these beaches. When the industrial and municipal polluters complained about the cost of an accelerated cleanup schedule, federal authorities replied that failure to reopen the beaches might touch off race riots. Such riots would be directly attributable to the withdrawal of recreation facilities because of the pollution, and the blame would rest squarely on the shoulders of the polluters. Whatever the material costs that might have resulted from the damage and the additional costs necessitated by travel expenditures to other swimming sites, and the building of substitute swimming pools, the polluters apparently agreed that there were also political and public relations costs involved which more than justified expensive remedies. The accelerated cleanup program was accepted. Yet is this also the appropriate measure to apply to the estimate for cleaning up Lake Erie? Cleveland's civil rights problems may not be as directly affected by pollution in Lake Erie, but certainly there is something equally tragic in watching the death of the fish industry and the cessation of recreational swimming in one of the Great Lakes most accessible to large numbers of people.

Other costs of water pollution should include estimates of the damage suffered by the fish industry. The shellfish industry is claimed to have suffered a $45 million loss because water pollution in tideland areas led to the spread of hepatitis through the clam and oyster beds. Estimates can also be collected of the losses suffered as a result of the destruction of sword- and tunafish arising from the mercury poisoning scare of 1970 and 1971. Tons of such fish were ordered withheld from sale and untold monetary losses resulted from price reductions caused by customers' avoidance of the uncontaminated fish on the shelves.

Many other estimates of cost must be handled in an equally arbitrary fashion. What value should be placed on the premature death of an engineer from asthma induced by polluted air? Is a retired engineer who dies worth as much? What is the cost to society of "blue babies" who suffer from methemoglobinemia? This results when oxygen is boiled off from polluted water used to make baby formulas so that what remains contains increased quantities of nitrogen, which has an adverse effect on the stomach and bladder. What is the cost of mercury poisoning? How does one determine the value of the 116 Japanese citizens who were paralyzed or killed by mercury poisoning at Minamata? In addition to the expenses of medical care, should we also calculate the loss in earning power of pollution victims in the years to come?

Somewhat less tenuous but still arbitrary are the estimates of the losses due to plant closings or relocations because of pollution. Even more questionable are the estimates of the losses from industries which decide not to locate in an area because of pollution. For places like New York City, these are vital but obviously difficult questions to evaluate. Estimates of this nature have been made to show that the benefit to Pittsburgh of clean air would offset the costs involved in trying to clean it. Whatever losses due to industries that left Pittsburgh because they could no longer use the skies as sewers were more than offset by other industries and institutions newly attracted to or contented to remain in a cleaner Pittsburgh. Today Pittsburgh is a city with a revitalized economy. Moreover what was once America's smokiest city now has air quality that is comparable to most American cities of its size.

A specific case of air pollution's effect on the costs of operation of a firm is illustrated by what happened to the companies that fill air tanks for scuba divers in New York City. Public health officials suddenly realized that the air tank companies were compressing polluted New York City air into their tanks. It was bad enough for those who have to breathe New York air in more relaxed circumstances, but to have to breathe it in concentrated form several feet under water was obviously dangerous. Accordingly, the New York City Air Pollution Control Act of May, 1966, prohibits the sale or distribution of compressed air tanks for underwater breathing use without special permits. The cost of air pollution in this case could be measured by ascertaining how many compressed air companies have been forced to close their doors and how much existing firms have had to spend in new equipment necessitated by the new law.

In the same category of cost estimates is the loss to the optical industry that comes from the difficulty of selling contact lenses in a city like New York. Doctors there have found that it is extremely dangerous for their patients to wear contact lenses for any length of time because of the possibility of lacerations to the eye. Specks of dirt and ash are periodically caught between the retina and the lens. Presumably the loss entailed from this could be calculated by determining the per capita use of contact lenses in other less polluted cities and then comparing this with similar ratios in New York City. To this should also be added the cost of eye surgery necessitated by pollution.

Attempts have also been made to evaluate any additional living costs for a family in an air-polluted area. In one of the few systematic studies ever made, Irving T. Michelson of Consumer's Union in New York City estimated that a family owning its own home in New York paid over $800 (or $200 per person) more a year than a family located in a less polluted area. He estimated that families living in apartments pay on the average of over $400 more. This includes added expenditures for household maintenance—extra painting, cleaning and washing, extra laundry bills—and extra expenses on personal cleanliness. It does not include an estimate for extra medical expenses, replacement of clothing, fewer sunlight hours, and lower real estate values.

Michelson's figures may be a little high compared to $65, the figure most commonly cited as the cost of air pollution per capita in the United States. Unfortunately no one really knows whether Michelson is high, low or in the middle. One of the most embarrassing acknowledgements that economists studying environmental disruption should make (and often don't) is that so few studies have been made of the cost of air or water pollution. Some economists have tried to determine the effect of air pollution on land values, and one or two pioneering efforts to determine costs have been made, but the results so far suggest that no one need worry that the subject has been exhausted.[5]

The method used to arrive at the more generally accepted figure of $65 illustrates the flimsiness of past research in this area. As explained in Part II, the $65 figure originated in Pittsburgh in 1913, when a group of researchers roughly calculated how much air pollution added to their yearly living expenses. The figure they arrived at was $20 per person. Subsequently this was multiplied by the change in the cost of living index to bring it up to 1959 price levels. Then it was multiplied by the size of the American population to produce an estimate that air pollution costs the United States $11 billion a year. Analysis of air pollution in the

[5]Robert E. Kohn, "Evaluating Air Quality Standards by Comparing Abatement and Damage Trade-Offs Between Pollutants," mimeographed, paper presented at Winter Meeting of the Econometric Society, December 28, 1969; Robert J. Anderson, Jr. and Thomas D. Crocker, "Air Pollution and Residential Property Values," mimeographed, paper presented at Winter Meetings of the Econometric Society, December 28, 1969.

United States must fall back on such imprecise scientific estimates because nothing else is available. More recently the National Air Pollution Control Administration has published cost calculations that range from $14 to $18 billion. But in the light of the way such figures have been derived in the past, this may be nothing more than the $11 billion for 1958 adjusted for inflation and population growth to 1970. Nonetheless the fact remains that the costs of air pollution are high.

As uncertain as they are, the estimates of losses endured because of environmental disruption are not much better than estimates of the cost to *eliminate* environmental disruption. Almost all the same statistical hazards also apply here, and there are also several extra pitfalls. First, there is confusion between the costs required to construct an adequate treatment system and the costs of operating that system—often the figures are used without making any distinction. Second, it is not always clear just what the elimination of environmental disruption would mean. Thus when speaking of water treatment, is the goal secondary treatment or the more elaborate tertiary treatment? Primary treatment of one thousand gallons of water with removal of one-third of the Biochemical Oxygen Demand (BOD) costs about 3–4¢ per thousand gallons.[6] (For a definition of BOD and an explanation of what is involved in the various stages of treatment, see the Goldman and Shoop article in Part I). Secondary and primary treatment of a similar quantity of water, in which 90 percent of the BOD and about one-third of the nitrogen and one-third of the phosphates are removed, costs about 15–20¢. Another 15–20¢ must be added for the tertiary treatment of water, so that total costs amount to 30–40¢ per thousand gallons.[7] But we are a long way from reaching the goal of even secondary treatment in the United States. Currently the homes of 70 percent of our population are served by some kind of sewer system. However the sewage from about 7 percent of these homes (involving 10 million people) is discharged directly into our water courses as raw sewage. Only about 43 percent of the American population (about 85 million people) is served by secondary sewage systems.

For many people living in isolated areas there is no need for secondary treatment or even for sewers. A septic tank or even an outhouse may adequately treat such waste if the site is isolated enough. (In effect, this is nature's way.) As people continue to crowd into urban areas, how-

[6]S. Fred Singer, "Environmental Quality; When Does Growth Become Too Expensive?" mimeographed paper presented in the Symposium "Is There an Optimum Level of Population?" American Association for the Advancement of Science, Boston, Mass., December 1969, p. 13.

[7]*Ibid.;* Gene Bylinsky, "The Limited War on Water Pollution," *Fortune,* February 1970, pp. 104–5, 194.

ever, the need for sewers and secondary treatment grows. The Federal Water Pollution Control Agency (now the Water Quality Office) estimates that by 1974, 90 percent of our urban population will need secondary sewage systems. To bring us up to that level, expenditures of over $10 billion in water treatment plants (exclusive of land costs) and over $6 billion in sewers will be required during the years 1970–74.[8] To separate storm and household sewers could cost anywhere from $15 to $50 billion.[9] Construction of secondary treatment facilities for industrial wastes will require close to another $5 billion, and control of sediment and acid drainage from mines could require from $2 to $5 billion. To control thermal pollution (overheating the water by discharging water used for cooling) will cost yet another $2 billion although heightened standards may soon necessitate a figure as large as $4 billion. This would bring the total investment required for industry to something like $9 to $14 billion. Total capital requirements for secondary treatment of water therefore will probably range between $40 and $80 billion. The Federal Water Pollution Control Administration estimates that annual operating costs for all these facilities would reach almost $2 billion for municipal plants, $3.5 billion for industrial plants, and about $1 billion for thermal processes.

If we were forced to move to tertiary treatment, the total for all construction costs in Table II would probably jump from $40 billion to about $90 billion. Even if very few American cities decide to move to tertiary treatment and only a few spend the money to separate storm and household sewers, it is still easy to see why some authorities estimate the ultimate cost at close to $100 billion.[10]

It is necessary to remember that not all of this expense is incurred at the treatment end of the process. Large sums of money will also have to be spent periodically by manufacturers to alter their product mixes to make them less destructive to the environment. For example, it is calculated that the shift to soft or biodegradable detergents cost the chemical and soap industry over $100 million. Such a move was ordered in Germany in October 1964, and later adopted by individual American states like Wisconsin because more and more sewage plants found they

---

[8]Federal Water Pollution Control Administration, U.S. Department of the Interior, *The Cost of Clean Water,* Summary Report, Vol. I (Washington, D.C.: Government Printing Office, March 1970), pp. 5–8.

[9]A system of holding tanks for the temporary storage of storm water overflows might obviate the need to construct separate sewers. This would cost a mere $15 billion. (See *Ibid.;* also *New York Times,* February 25, 1970, p. 59; *Fortune,* February 1970, p. 195.)

[10]*New York Times,* March 17, 17, 1970, p. 29. CEQ *Environmental Quality,* Second Annual Report (Washington, D. C.: Government Printing Office, August 1971) pp. 111, 123. The CEQ's cost estimates for 1970–75 are slightly lower than ours would be if extended to 1975.

could not eliminate the growing sea of suds in their sewage. Finally in July 1965, the whole American chemical industry started to make detergents out of linear alkylate sulfonite instead of the hard alkyl benzine sulfonite, that could not be decomposed easily in most sewage treatment plants. About five years later the soap manufacturers were faced with the need to spend millions of dollars more to make another set of changes. This time the focus was on the reduction of phosphates in the detergents. The usual secondary treatment system is unable to remove sufficient quantities of phosphates from the water. Since phosphates are also used as fertilizer, the discharge of large quantities of phosphates into the water has had the effect of stimulating the rapid growth of aquatic plant life. This has increased the spread of algae and seaweed, which in turn eat up the oxygen in the water, hasten the formation of swamps, and accelerate the demise of the water course. (Technically this is called eutrophication. What is happening to Lake Erie is a prime example.) While one solution to this problem would be to build tertiary treatment plants across the country, the cheaper approach is elimination or reduction of the phos-

Table II: The Costs of Air, Water and Solid Pollution Control for the Years 1970–1974 (in billions of dollars)

	Total Construction Costs	Annual Operating Costs
A. Water Treatment Costs (Secondary Treatment)		
1. Household sewage		
a. Treatment plants	$10	
b. Sewers	$ 6	
	$16	$2
2. Sewer separation	$16	
3. Industry	$15–$50	
a. Industry treatment costs	$ 5	$3.5
b. Sediment & acid drainage costs	$ 2–$5	
c. Thermal pollution	$ 2–$4	$1
	$ 9–$14	
Total Secondary Treatment	$40–$80	$6.5
Tertiary Treatment	$90–$100	
B. Air Pollution		
1. Stationary sources	$ 8–$10 ($50–$100)*	$.3–$2
2. Automobile pollution	$ 1–$ 2	$2–$3.5
C. Solid waste	$ 1–$ 3	$3–$8
Total Pollution Control (Secondary Water Treatment only)	$50–$95	$12–$20

*The higher figure is for the year 2000.

phates at the soap factory before they even enter the sewers. Nonetheless the search for a suitable substitute and the alteration of the production process will seriously affect an industry which does a business of about $4 billion a year. Similar examples of how money for pollution control has been spent before the product was produced, rather than on the treatment of the resulting effluent, can be found in the production of paper, where manufacturers stopped using the sulphite process and switched to the sulphate process.

Another way to calculate the cost of eliminating water pollution is to see how much it would cost to manufacture fresh water. This presumably would represent the outer limit of possible costs, but in some areas reprocessed water is the only alternative source available. Therefore the cost of such reprocessing might serve at least as the upper limit for the value of the water, even though such estimates do not measure the cost of pollution directly. This may be a fairer estimate of the cost of polluting the water, since the practice otherwise is to set the price of fresh water at the amount necessary to cover only the operating costs of the well. Such a system fails to make allowance for the fact that the well may be slowly going dry, as is often the case in the West.

Traditionally attention has been focused on the cost of refining sea water. Given present technology, this is quite expensive. The best desalinization plants can almost produce drinking water at a cost of one dollar per thousand gallons. The facility at the United States Naval Base in Guantanamo, Cuba, produces one thousand gallons for $1.16. This is a considerable improvement over the $14 per thousand gallons that water desalinization cost in 1952. At one time it was hoped that with the aid of special subsidies, an atomic-powered desalinization plant could be built to serve Los Angeles. It was assumed that the subsidies, combined with the revenue from the sale of electric power, would make it possible to price water at about 27¢ per thousand gallons. In late 1970, E. I. du Pont de Nemours & Co. introduced a reverse osmosis system that costs from 25–65¢ per thousand gallons depending on the salinity of the water and the scale of the desalinization plant. Yet unless special subsidies are provided, as was contemplated in Los Angeles, it is unlikely that desalinated water will be cheap enough to compete with the normal cost of well or reservoir water which averages about 15¢ per thousand gallons. However if the cost of sewage treatment is added in, the total cost is closer to about 23¢ per thousand gallons. Some authorities have argued that this cost disparity between fresh water and desalinated water will always exist because scientists are misdirecting their attention toward desalinization. Instead such critics urge that more effort be devoted to reusing sewage water. Sewage often has only 1 percent impurities while sea water has about 3 percent impurities and therefore is harder to clean.

There have been two or three experiments with reprocessing sewage water. At Lebanon, Ohio, a treatment plant operated in conjunction with the Taft Center of the United States Public Health Service has been able to reprocess drinking water at a cost of 54¢ per thousand gallons. This is one-half the price of similarly treated salt water and could be very promising for the future provided that engineers can overcome the aesthetic resistance that people have toward drinking reprocessed sewage. An even more promising demonstration has been conducted at Windhoek in the dry highlands of South Africa. There at a cost of about 30–40¢ per thousand gallons, sewage or brackish water is subjected to such thorough tertiary treatment that it is then used for drinking.[11]

A related problem is the disposal of sewage water. In some areas there is no convenient body of water that can be used as a dumping ground for a city's sewage. This is the problem at Lake Tahoe. If the lake, one of nature's purest, is not to be used as a sewer itself, some other repository must be found. Unfortunately there are no other outlets available in the area so plans have been drawn up to pump the sewage several miles over the mountains. Conservationists in the USSR have argued for a similar arrangement to dispose of effluent from the paper plants on Lake Baikal, but to no avail. The Russian authorities simply argue that it is too expensive.

Santee, California, also had a disposal problem. A treatment plant was built and the water was then further purified by being run over a dried-up riverbed. After a mile of this natural filtration, the water was stored in small artificial lakes stocked with fish and opened to boating. Although unfit for drinking, some of the water is being sold to golf courses for watering the grass. The price is 12¢ per thousand gallons, which is cheaper than fresh water. Since some nutrients are probably still left in the water, this may actually turn out to be a bargain since the water can also be used as a fertilizer.

Estimates of eliminating air pollution are even more complicated. Projected operating costs range from $300 million a year to $2 billion and capital costs over the five years from $8 to $10 billion.[12] According to the maximum forecast, that would mean capital expenditures of slightly less than $100 billion by the year 2000. Twenty billion dollars is probably a much more reasonable figure. But these cost figures only cover emissions

[11]Singer, *Environmental Quality*, p. 13.

[12]Committee on Public Works, United States Senate, *On Pollution*, 1967. (Air Quality Act) S. 780, Part 3 (Washington, D.C.: Government Printing Office), p. 1361; *Fortune*, February 1970, p. 123; James J. Hanks and Harold D. Kube, "Industry Action to Combat Pollution," *Harvard Business Review*, September–October 1966, p. 57; The Working Committee on Economic Incentives, *Cost Sharing With Industry?* (Revised) (Washington, D.C.: Room 325, Executive Office Building, November 20, 1967), p. 21. CEQ Second Annual Report, p. 111.

from stationary sources. Currently, the greatest source of air pollution in many cities is the automobile. A variety of solutions for reducing emissions from the internal combustion engine have been suggested. They include everything from outlawing the combustion engine itself to eliminating gasoline with lead. The cost of the cure varies with the severity of the remedy. There seems to be some agreement (this may merely be a sign that everyone is repeating the original estimate) that proper control equipment and the elimination of leaded gasoline would result in a 10 percent yearly increase in the cost of operating an automobile. The cost per car would be about $30 a year based on the average per capita expenditure of $300 a year that Americans spend on their automobiles.[13] For the country as a whole, the total annual outlay could range from $2 to $3.5 billion.

Another cause of air pollution is the jet airplane. Generally airlines claim that they contribute only 1 percent of the nation's air pollution. However, the amount of jet pollution varies from city to city and neighborhood to neighborhood, depending on proximity to the airport.[14] New York City reportedly receives one and a half tons of air pollutants a day. As a result of new laws instituted by officials in New Jersey and Chicago, which now allow state officials to exact fines as high as $5,000 for planes which violate exhaust standards, the airlines have agreed to curb their exhaust emission by 1972. Initially airline officials claimed this would cost $30 million. Subsequently they lowered their estimates to $13–$15 million.

Although it is not indicated in Table II, it is necessary to remember that some of the cost incurred in reducing water and air emission will be offset by savings that come with cleaner air and water. One benefit of cleaner air would be a reduction in cleaning bills and medical expenses. Another saving will result from increased utilization of the by-products of combustion. This will permit greater recycling of nature's resources and will mean less exploitation of new resources. It is estimated that approximately 23 million tons of sulfur dioxide are discarded into the air each year as part of the combustion process.[15] If recovered, this could provide about 5 million tons of sulfur or 15 million tons of sulfuric acid. Although the price of sulfur has fallen in recent years so that recovery is less attractive than it once was, such a program of resource reuse would virtually satisfy the entire demand for some sulfur compounds.

Institution of $50 to $100 billion worth of air quality controls over the next few decades would not mean the elimination of all air pollution, nor of the costs that arise from it. Nevertheless, substantial savings could be realized and therefore used to defray the cost of the annual expenses

[13]Resources for the Future, Inc. *Resources,* no. 33, January 1970, p. 8.
[14]*New York Times,* January 21, 1970, p. 26.
[15]Hanks and Kube, "Industry Action to Combat Pollution," p. 54.

envisaged as necessary to clean up the air. This could go a long way if not all the way toward offsetting the annual $5 to $13.5 billion in operating expenses that would be needed to curb air pollution as indicated in Table II.

The cost estimates for solid waste disposal are somewhat less complicated but still something less than precise. They are less complicated because in the present state of the art, the technology is even more primitive than it is in water and air treatment. Usually solid waste disposal involves nothing more than an incinerator and often just an empty field and a fleet of dump trucks. Sophistication may mean nothing more than an automobile shredder which may cost $3 million. Under the circumstances, the capital expenditures required for solid waste disposal are negligible compared to the investment required for air and water treatment. Thus to dispose of the estimated 5 pounds a day in household waste discarded by an average American, the major costs are collection, transportation and disposal. It is estimated this amounts to from $3 billion to $8 billion a year.[16]

It is unlikely that the exaggerations have balanced the understatements of these various estimates. Consequently combining the sums of the different construction projects and operating costs makes it doubtful that the uncertainties stressed above have been eliminated. Having gone this far however, we might just as well add them together to have a rough picture of how much they would all cost. The overall total of the different projects indicates that reducing environmental disruption could cost anywhere from about $50 billion to something approaching $100 billion. Annual operating costs would probably range from about $12 to $20 billion. These operating costs amount to approximately 1 to 2 percent of the annual GNP and 4 to 7 percent of the value of our industrial, agricultural, mining, and transportation output. Construction costs average about 3 to 7 percent for industry but may total more than 10 percent of the capital costs of some projects.

It is necessary to reemphasize that all such calculations are imprecise. Even if it is assumed that we know what standards we want to attain, it is still hard to project meaningful figures. Treatment technology is as yet relatively primitive. Technological breakthroughs could reduce costs significantly. Moreover, the estimates will be affected by what happens to population growth and concentration. If population increases, so will costs. It will also depend on what new industrial processes complicate existing treatment methods. Care must also be exercised in judging cost estimates because there is a tendency to overstate costs. Sometimes this is done by those who feel that the only way to obtain action is to

---

[16]*New York Times,* March 27, 1970, p. 28; CEQ Second Annual Report, p. 111.

create a crisis atmosphere. Regrettably, too often it takes a scandal or crisis before public support for meaningful programs can be obtained. At other times the costs are exaggerated for just the opposite reasons: to discourage pressure for change and also to win admiration for the expenditures already made. This is understandable, since almost everyone agrees that pollution control expenditures usually add nothing to profits. On the contrary, they generally result in diminished profits for industry (except for the producers of pollution control equipment) and in higher taxes.

Finally, it is difficult to determine how much man must do to clean up his water and air. Up to a certain point, water courses and air sheds have a natural self-cleaning power. This regenerative process has served admirably for the last several thousand years. It breaks down, however, when a threshold is passed and the amount of newly added dirty water and air suddenly overtaxes nature's cleansing capacity. The question that arises is how much should we expect nature to do on its own and how much should we require man to spend to facilitate the process? In some cases, man's help is a necessity; in other cases it is unnecessary. Deciding where nature's work ends and man's work begins is not easy.

One of the greatest challenges open to an economist today is to make a careful study of the effects of pollution. Unfortunately neither exaggeration nor understatement will ordinarily advance the long-run cause of those who seek reliable economic data. Until there are reliable estimates of damage, no appropriate value can be attributed to clean air and water. And, as long as the cost of clean air and water is undervalued, they will be used wastefully and irrationally.

Even if better economic data on costs of environmental disruption were available, the impact of the costs of such a pollution control program on the GNP would still be hard to determine. In large part the calculation is dependent on the assumptions that are made and how one defines GNP. How, economists like Walter Heller and others ask, can you count the sales value of a chemical plant's output if in the process of production, fish are killed and the fish catch is reduced?[17] In the same way, how can you count the output of machinery for chemical waste treatment when this machinery really does not add to production; it merely does what should have been done in the first place? A resource has been consumed in the process of production and this should be recognized by subtracting the loss due to pollution from the value of the company's output.

[17]Walter W. Heller, *New Dimensions of Political Economy* (Cambridge: Harvard University Press, 1966), p. 111; Shigeto Tsuru, ed., *Proceedings of International Symposium on Environmental Disruption: A Challenge to Social Scientists* (Tokyo: Asahi Evening News, 1970), p. 56; Shigeto Tsuru, "In Place of GNP," mimeographed, January 1971.

On a national basis, this could be reflected in the substitution of a concept like Net National Well-Being (NNWB) for Gross National Product (GNP). In such a system the produced value of items like steel and chemicals would be counted net of the pollution that is generated in the course of production.

Under present practices, expenditures for pollution control do count as additions to GNP. Consequently, an energetic enforcement program of pollution control would probably not result in a significant fall in GNP as presently defined, especially in the short run. Expenditures and incomes would merely shift somewhat from consumption industries to pollution control industries as indicated earlier. However, the dynamic effect of such a change is a little harder to predict. A reduction of consumption relative to the rest of GNP could affect incentives and lead to a fall in GNP just as an increase in consumption relative to total GNP led to a jump in the GNP. For example, when the income tax rates were cut in 1964, the increase in consumption set off a spurt in the growth of GNP. The fall in consumption due to the diversion of resources for pollution control could have the opposite effect and this could precipitate a fall in GNP.

Some critics have argued that the only long-run solution to environmental disruption is to call a halt to industrialization and economic growth. They argue that it is our fetish with growth that has caused such abuse to nature. To some extent they are right. Industrialization is like cooking an omelette. There is no way to avoid cracking eggs. Not only is this hard on the egg, but we then have the shells to dispose of. Yet it is unrealistic to expect that many people will agree to forego their expectations of increased comforts, much less give up existing comforts.

POSSIBLE SOLUTIONS OR
HOW TO CLEAN UP THE MESS

Having outlined the problem and having shown how the existing economic stimuli tend to encourage pollution, it is now necessary to see what can be done to reduce and eliminate it. In other words, what are the ways to cope with external diseconomies? How can social and private benefits and costs be brought into closer conformity?

There are at least three ways to solve the problem: 1. Direct Regulation; 2. Subsidization; 3. Incentive Pricing. An efficient solution would undoubtedly involve all three techniques. But before we can discuss these various methods in detail, it is necessary to decide what our goals should be.

How clean should we make the water and air? There is no widespread agreement about this. Even if it were possible to measure pollution, authorities would probably differ widely over standards for water and air. Many conservationists for example refuse to discuss whether some streams and air spaces should be designated as sewage receptacles They insist that no pollution whatsoever be permitted. To most economists this is an unrealistic approach, since some waste in an industrial society is inevitable. The cost of eliminating all waste might necessitate the cessation of production. As economists like to say, the costs of control may well exceed the benefits. This means that money should be spent on pollution control only so far as the cost of further abatement does not exceed the benefits obtained from such measures. Others suggest that a classification system be set up as New England water authorities have done and as now adopted in one form or another by the Federal Water Quality Administration across the nation. As described in Part II, New England and now federal officials have designated certain uses for different water courses. It is a bit like real estate zoning—certain rivers and water bodies may be used for industrial purposes (dumping wastes) while others are to be preserved strictly for residential uses (drinking and swimming). Essentially the same kind of scheme has been adopted for air control.

But assuming that some standards can be agreed upon, how are they to be implemented and enforced? One of the difficulties with the zonal system is that there is a tendency to pull the water standards down to the next level of classification. Someone always has a good argument for making the water just a little dirtier.

Direct Regulation

The traditional approach to pollution has been legal regulation. When faced with a rolling tide of polluted air and water, the instinctive reaction of most government agencies is to pass a law ordering the dirty water and air to halt. But as King Canute found, command does not necessarily assure compliance.

One of the difficulties with pollution laws is that often they are too sweeping in nature. Frequently laws will be enacted that demand an inappropriate standard of pollution control or fail to provide for the financial burdens which such laws entail. Experience seems to show that laws will be ignored or violated if they involve substantial inconvenience or widespread economic loss for too many individuals or groups. Violations will be especially blatant if there are inadequate detection and policing systems. Some automatic or self-regulating enforcement mechanism is necessary if the laws are going to be observed. In the absence of adequate inspection and enforcement, antipollution laws will not be adhered to until they provide direct benefits for the potential violators offsetting the expense of compliance. When policemen are not around, traffic laws are usually observed even though this means a delay, because failure to do so may lead to automobile accidents. In contrast, the Volstead Act could not be enforced because too many citizens felt that the consequence of alcohol was pleasure rather than pain.

Observance of pollution laws is especially difficult when it involves financial sacrifice. In some cases, an inevitable result of such laws is that some industries must terminate their production. At the very least, the installation of pollution control equipment will mean increased costs of production and lower profits. The economic losses necessitated by pollution control are usually weighed against the possibility of having to pay a legal fine for violation of the law. Generally the fines are so low, the laws so poorly enforced, and the costs of pollution control so high, that polluters frequently prefer to continue their past violations.

State and local laws.   One of the reasons for the lack of compliance with most pollution laws is that such laws traditionally have been passed and administered at the local and state levels. This means that over the years there is little or no enthusiasm in their administration or enforcement. Having passed well-intentioned legislation against pollution, most lawmakers consider their jobs to be done without providing for the inspection, enforcement, or financing of those efforts. Furthermore, state and local agencies in the past have often found themselves susceptible to various kinds of economic and political pressure from those against whom the laws are directed.

Massachusetts has had considerable difficulty enforcing a set of potentially effective pollution laws even though theoretically it has been a leader in water pollution control. It established the first Department of Public Health at the state level in 1869. In 1886, it set up a laboratory for pollution control on the Merrimack River at Lawrence. Simultaneously, a strict pollution control law was enacted, one of the first in the country. The Merrimack River at the time was one of the worst industrial and municipal sewers in the country. Because it was a source of power as well as water, some of the earliest American industries—textile, leather and paper—were located along its banks. Consequently the river communities around Lawrence and Lowell constituted a mecca for three of the worst types of industrial polluters in the nineteenth century. It did not take much insight to recognize that the existence of a pollution laboratory along with potentially effective laws posed a threat to the economic well-being of one of the country's industrial centers. In response the industrial interests in the region sponsored a second law that was passed in 1887, the year after the laboratory had been approved. This latter law exempted the Merrimack River from pollution control, an exemption that remained in effect until 1946.

Only by the mid 1960s was any significant effort made to prevent Lowell and Lawrence from dumping raw sewage into the river. By then, most of the textile, leather, and paper factories had moved south. Consequently the threat of economic loss from industrial relocation or renovation due to pollution control no longer intimidated those who were in favor of pollution control. Also, with time, it had become apparent that failure to eliminate pollution entailed its own costs. Other industries were avoiding the area because of the pollution. Maine has had similar problems and many of its rivers have been affected by the same three industries. Yet because so much of Maine is still underdeveloped, its land and labor are cheap and there are still large quantities of untapped natural resources available. As a result, Maine is regarded as virgin territory by promoters of oil refineries, aluminum smelters, and atomic power plants. To many underemployed residents of the state, the promise of all the new jobs such industry would bring has been very tempting. To other local residents, such industry would destroy the last-frontier character of the area. They were supported by out-of-state vacationers who made their money and generated their effluent in larger metropolitan areas to the south. With their money, they sought to flee their own polluted areas in order to seek the natural beauty of Maine. Inevitably many of the poorer permanent Maine residents regarded this as a hypocritical attempt to preserve nature at their expense. In other words, it was all right to make money and pollute in Boston but not in Machiasport, Maine.

In an effort to balance these opposing interests Maine's legislature created this nation's first state Environmental Improvement Commission in 1970. The location of all prospective industries on sites larger than twenty acres which might substantially affect the environment must now meet the approval of this group. At the same time it has been given extensive power to regulate the storage, processing, and shipping of oil along Maine's coast. Although the power of such organizations is being tested in the courts, Maine hopes through such controls to be able to attract and locate industries so that they will not detract more than they add. The citizens of Maine have come to realize that a polluting factory may turn away more business than it generates in payrolls and production. Vacationers may avoid the area and higher taxes may be needed to provide public money for pollution control.

As a state develops a varied economic base it is less subject to the pressures of individual industries that find a program of pollution control to be too costly. This trend toward less domination by industry is likely to continue as services grow in importance in the economy and agriculture and industry decline. Due to such tendencies, states find themselves mediating between more varied interest groups, including an ever larger and more outspoken conservation sector. Organizations such as the League of Women Voters and the Sierra Club have been particularly effective in forcing state officials to take a stance independent of particular industrial interests. Consequently a growing number of states like Massachusetts, California and New York have belatedly started to introduce effective legislation to control water and air pollution. Nonetheless the most effective form of direct legislation has been initiated at higher political levels.

Interstate control. One of the obstacles to effective legislation for air and water pollution is that most of our main water and air basins overlap several states. Our cities and states do not always encompass complete air and water systems. Therefore even when one local unit eventually decides to take effective action, its efforts may be nullified by its inability to induce similar action by adjacent governmental units in the basin. Consequently any control organization must make provision for the overlapping of political boundaries and the transfer of adequate authority and funds to new and specially created groups.

Examples of the difficulties that exist are illustrated by what happened in the attempts to clean up the Raritan, Delaware and Ohio River basins (see Part III). Similarly, when the federal government sponsored a conference of all interested parties to prepare for the cleanup of Lake Erie, the governors of New York and Pennsylvania withdrew their representatives temporarily because they felt the pollution of Lake Erie was a matter for the states to solve individually. New York City has encountered similar obstacles in its efforts to clean up its air. It has been unable to obtain compliance with its rules from polluters in New Jersey. Sometimes special interstate compacts are necessary to authorize the formation of suprastate organizations. However because of their unique nature, such organizations are not always effective.

Federal control. Jurisdictional problems are not limited to the regional level. There are countless difficulties at the federal level itself. For that matter, international jurisdiction is sometimes disputed, as seen in this country's argument with Canada over responsibility for Lake Erie, and with Mexico over responsibility for the Rio Grande. The evolution of controls within the federal government tells its own story. Basic responsibility for water pollution was originally assigned to the United States Department of Public Health when it was created in 1912. Prior to that, Congress had passed a Refuse Act in 1899 that prohibited the discharge of refuse into navigable waters. However because enforcement of the Refuse Act was the responsibility of the Secretary of the Army and no full-time agency was created to administer this law, it had little effect. Subsequently the Oil Pollution Act of 1924 was enacted to prohibit the discharge of oil into coastal water. This too was administered by the Secretary of the Army, mainly through the Corps of Engineers.

Primary responsibility however for pollution control continued to fall on the Department of Public Health. Its responsibilities were increased after the passage of two new and more effective laws, the Federal Water Pollution Control Acts of 1948 and 1956. But the jurisdictional problems began to grow when the Public Health Service was incorporated into the Department of Health, Education and Welfare when it was formed in 1953. At that time the Department of Health, Education and Welfare was also responsible for air pollution control. Then in 1965 the Water Quality Act created the Water Pollution Control Administration.[18] There had been considerable feeling that the Public Health Service had not taken a strong enough stand against industrial pollution. Shortly after its formation, the Water Pollution Control Administration was transferred out of the Department of Health, Education and Welfare and into the Department of the Interior, which had traditionally held responsibility for water resources. To show that they were thinking positively, officials in the Department of the Interior then changed the name of the Federal Water Pollution Control Administration to the Federal Water Quality Administration. In 1971 it was changed again to Water Quality Office and then to Office of Water Programs. If only it were as easy to change the environment!

Neither the alteration in name nor the switch to the Department of the Interior did much to eliminate the duplication and division of responsibility. For a time conditions were actually worse. Whereas at one time the Department of Health, Education and Welfare administered control over both air and water agencies, after 1965 it was left to administer only the air pollution laws. At various times some form of environmental responsibility has also been delegated to agencies in such organizations as the Department of Housing and Urban Development, the Department of Agriculture, the Department of Commerce, and the Department of Defense.

Fortunately before too much empire building and duplication could

[18]For an enumeration of the legislation on water quality from 1899 to 1969 see J. Clarence Davies, III, *The Politics of Pollution* (New York: Pegasus, 1970), pp. 38–49; CEQ First Annual Report, pp. 43–5.

take place, officials in both the administration and Congress recognized their previous shortsightedness. Consequently in late 1970, the Environmental Protection Agency (EPA) was formed with William D. Ruckelshaus as its director. Within its doors have been gathered the functions previously supervised by fifteen diverse organizations with 5,800 employees and budgets of $1.4 billion. For example, EPA took the water pollution activities (Office of Water Programs) from the Department of the Interior, air (Office of Air Programs) and solid waste control (Office of Solid Waste Management Programs) from the Department of Health, Education and Welfare, supervision of pesticides (Office of Pesticide Programs) from the Department of Agriculture, establishment of environmental radiation standards from the Federal Radiation Council (Office of Radiation Programs) and the enforcement of those standards from the Atomic Energy Commission. It will also assume from the Army Corps of Engineers the power to issue permits for the discharge of materials into the oceans and Great Lakes. Still, the Army Corps of Engineers will retain its authority over the issuance of permits for dumping in navigable inland lakes and streams.

The decision to allow the Army Corps of Engineers to keep its fingers in the pollution pie suggests that this newest reorganization will not eliminate all duplication and dispersion of authority. This is further indicated by the fact that at the same time the EPA was set up, President Nixon also decided to create the National Oceanic and Atmospheric Administration. This too absorbed functions that previously were held by units of the Department of the Interior, the National Science Foundation, the Coast Guard, the Navy and the Army Corps of Engineers. With a budget of $270 million and a staff of over 12,000, this organization seems to water down the promise of unification that the EPA presumably was to have provided. Undoubtedly other pockets of control have also been missed by the EPA. Nonetheless there certainly has been considerable consolidation of administrative responsibility for environmental affairs. The formation of this cabinet-like agency means the fight against environmental disruption will be much better coordinated in the future. President Nixon also created the White House Council on Environmental Quality (CEQ) as part of the National Environmental Policy Act of 1969. The CEQ, with a three-man council, is patterned after the Council of Economic Advisers and is intended to set general policy in the office of the President. In contrast, the EPA is more of an operating agency.

Now that some of the confusion over pollution jurisdiction within the federal bureaucracy has been reduced, the next step is to establish cleaner lines of authority between state and federal government. Federal authorities have been reluctant to interfere in cases of air and water pollution that affect only one state. It was only in 1969 that the federal government imposed its own water quality standards on a state. This came after Iowa rebuffed federal efforts to nudge the state into the decision to install secondary water treatment facilities on intra- as well as interstate water courses. But even on interstate pollution, the federal government has acted with considerable hesitancy for fear of usurping state and local pre-rogatives. Although Presidents Johnson and Nixon have called for more power for federal agencies to initiate and enforce remedial action on their own, most federal agencies are still weak when it comes to enforce-ment. In recognition of this, Congress enacted the previously mentioned Water Quality Act of 1965, which called upon the states to draw up their own standards for statewide water quality by June 30, 1967. If the states refused or the standards were deemed to be too low, then federal stand-ards would be imposed. The same procedures were adopted for the estab-lishment of air standards in the Air Quality Act of 1967, although a much tougher law the Clean Air Act of 1970, authorized the federal govern-ment to take a much stronger stand.[19] If need be, the federal government can now force the closing of industrial plants which pollute the air. The federal government under the 1970 law was also prohibited from enter-ing into contracts with any violator of the Clean Air Act.

Until the Iowa case, federal authorities always had to wait for an invitation from a state before they could take action to correct pollution affecting a single state. The Secretary of Health, Education and Welfare and the Secretary of the Interior, and now the Director of the EPA, did however have the power to initiate conferences in order to consider cases of interstate pollution and damage. Before issuing any order the federal government must hold hearings on the causes and proposed cures of pol-lution. A schedule of abatement procedures is then drawn up and a sec-ond conference is arranged six months later to see if the cleanup is going according to schedule. If there is no progress after another six months the matter is turned over to the United States Justice Department for prose-cution. Therefore a minimum of twelve months is required before there can be any effective enforcement. As of early 1971, fifty enforcement con-ferences on water pollution and ten on air pollution had been held, but only two or three resulted in court action. While some officials say that this indicates the success of the program, others argue that it only shows how timid the demands for cleanup under the traditional procedures have been.

Responding to the initiative of conservationist congressmen like Senator Philip A. Hart and Representatives Henry A. Reuss and Michael Harrington, the CEQ, EPA, and Justice Department decided to seek new

[19]*Ibid.*

ways to speed up the enforcement procedure. Digging to the bottom of the legislative barrel, they unearthed the original Refuse Act of 1899. Initially intended to facilitate navigation and prevent the dumping of objects that might jeopardize shipping, this law could also be wielded powerfully against the discharge of pollutants. Previously the Army Corps of Engineers had seldom used it for this purpose unless the flow of shipping was threatened. Potentially, however, they or the EPA could simply refuse to issue permits to dischargers unless the pollution hazards were eliminated. The advantage of this new approach is that no complicated hearings and compromises are needed to bring compliance. The Refuse Act provides for fines of $2,500 a day and criminal sanctions for any party —be it city, factory, or farm—that dumps wastes into navigable waters without a permit. Moreover since almost all water courses flow at one time or another into navigable waters, the potential for control is enormous. All that is required now is that previous permits be rewritten with the higher standards in mind. This is already being done. The flexibility of this old law was demonstrated when it was used so effectively to bring about the termination of mercury discharges into water bodies at the time it was discovered that mercury was accumulating so rapidly.

Partly in reaction to the mercury scare and to complaints that the federal government had not been doing enough to curb pollution, it was decided to utilize the old 1899 Refuse Act in an unprecedented way. Over 40,000 firms were ordered to apply to the Corps of Engineers before July 1, 1971, in order to secure validation of their old permits to dump. To continue their eligibility, all 40,000 firms had to obtain certification from state water pollution control agencies either that they were already complying with existing state and federal standards or were implementing an approved program that would lead to compliance. Each firm was to conduct an analysis of its effluent that must indicate fourteen of the most important water quality characteristics, such as the flow rate, dissolved solids, five day biological oxygen demand, and nitrate levels. Originally fifty-one other items, such as the level of radioactivity, mercury, DDT, and arsenic were also to be reported, but this requirement has been postponed for the time being.

In another return to the past, a group of Chicago lawyers belonging to Businessmen for the Public Interest dug out a federal law from 1910 that may prove to be even more effective than the 1899 Refuse Act. According to these lawyers, the 1899 Refuse Act is too lenient since it only requires that emitters conform to existing standards. The 1910 law, however, goes further in that it prohibits the discharge of refuse of any kind into Lake Michigan unless the discharge is contained behind a breakwater. If applied throughout the country, this regulation could be another powerful weapon in the hands of the Environmental Protection Agency.

Effective Laws. Despite considerable progress, a candid analysis of the effectiveness of the laws presently on the statute books shows that enforcement is often rather lax. Consolidation of federal administrative authority and the resurrection of the potentially powerful Refuse Act of 1899 and the law of 1910 will help, but many authorities complain something is lacking. In some cases enforcement has been hampered by intentional efforts to subvert the law; in other instances technology and industrial expansion simply moved too fast for control authorities to keep up with the latest developments. As a spur toward greater compliance with the laws, many conservationists have come to support the doctrine of the personal right to sue for damages inflicted on the environment. Led by Professor Joseph L. Sax of the Law School of the University of Michigan, the state of Michigan passed a law in 1970 which made it legal for anyone to sue in order to protect air, water, and other natural resources. The defendant may be an individual, a company, or the state. A citizen can therefore sue to prevent his neighbor from burning trash outdoors or to force a company to close down until it stops polluting. Previously the court ruled against such suits on the grounds that there was no legal precedent for them. Thus the residents of Santa Barbara were unable to sue the Department of the Interior to prevent it from granting leases for offshore drilling. If such a law were passed at the federal level, the oil companies and the Department of the Interior could have been ordered to halt such drilling.

Luckily, not all direct legislation is difficult to implement. Laws which stress prevention are usually more effective than those which insist on correction. Somehow it always seems to be easier to stop something before it starts than to eliminate it once it has begun. The polluter always hopes that existing pollution will be tolerated a bit longer so that the outlay of large sums of money can be postponed. For that reason, the use of the Refuse Act of 1899 should be particularly effective because prospective entrepreneurs now know that certain standards must be met before production even begins.

To the distress of many manufacturers and municipalities, however, the original standards are often altered, sometimes even during the course of construction. This is wasteful, since the economic loss entailed in correction is higher than if adjustments are required in the beginning.

Equally important for the success of direct legislation is an aroused public opinion. When compliance with a law entails an economic loss, there is more likelihood of remedial action if there is the added danger of customer or public disapproval. For example, after the public realized that the automobile was the main source of pollution in Los Angeles, it was possible to design a rigid law regulating exhaust control. Public acceptance of such a requirement is of utmost importance because its effects will be so widespread; observance of the law will mean that car buyers must directly pay for the new equipment. Since air pollution is not as serious in other parts of the country as in Los Angeles, the public elsewhere may *not* consider the extra expense for pollution control to be worthwhile. The standards set for 1975 for the whole nation may be particularly costly to achieve. Consequently a massive educational campaign may be required to generate the desired popular response.

An awakened public opinion was also important in securing support for air-pollution control in New York City. As the quality of the air in New York City deteriorated, the public became extremely outspoken about the resulting health hazards and aesthetic nuisance. Such sentiments have been effectively utilized by such groups as the New York Citizens for Clean Air, Inc., who have labored vigorously for more restrictive legislation. Similar public interest and action were harnessed on a national basis as part of the Air Quality Act of 1967. In some instances public participation apparently had little impact on raising the standards but in some areas public lobbying actually did result in more stringent codes.

Compliance with these laws necessitates expenditure of millions of dollars in new processing equipment and added fuel costs, but it sometimes happens that those who call loudest for an increase in pollution control are not prepared to pay the bill when it falls due. All too often the supporters of increased control seem to vanish when the time comes to support the increase in prices, utility rates, and taxes necessary to finance these controls. Yet despite the risk that financial support might not be forthcoming, some large fuel consumers like New York's Consolidated Edison will sometimes roll with the punch of public pressure and accelerate efforts to improve the environment. Since it would have to comply sooner or later with a law that required the use of fuel with a low sulfur content, Consolidated Edison voluntarily advanced the date of its use of low-sulfur fuels and was then in the admirable position of being able to boast that it had done its part. In fact its emission of sulfur dioxide fell by almost 25 percent from 1966 to mid-1969. Some skeptics discounted these efforts, pointing out that since Con Ed was a utility, buying low-sulfur but high-priced oil was not really such an altruistic gesture, since the extra cost could be immediately passed back to the electric consumer. Nonetheless most critics would still agree that Consolidated Edison is at least more positive than the National Coal Association, which ran the following ad a few years ago:

> If you want an instant end to air pollution . . . stop driving your car, then turn off your oil burner, brick up your fireplace, bundle your leaves, box your trash, refuse delivery of anything by truck, boycott airplanes, trains, busses and cabs. Don't use anything which requires oil, gas, coal or atomic energy in its manufacture—such as electricity, steel, cement, clothes, food, newspaper, babies' rattles and on and on and on and on . . . or let's face the fact that *any* combustion generates pollutants . . . and that any "instant end" to air pollution brakes our civilization to a halt. Coal is a minor cause of this contamination, but the coal industry is working hard to clean the air. After all, we're breathing it, too.

This is an atypical but dramatic example of industrial opposition to pollution control. It is largely a reflection of how serious the problem is for the coal industry, the source of a major portion of both our air and water pollution. Fortunately most American industry takes a more positive attitude, at least in public.

Regardless of how cooperative industry may be in adhering to new pollution regulations, it is often forgotten that industry is not always the only source of trouble. As it turned out, while Consolidated Edison complied with the new standards several months in advance of schedule, New York City itself has lagged. This lag is important because the city is a major air polluter. Before there can be a cleanup of the air, it will be necessary for the city to reequip its municipal incinerators, eliminate open burning in city dumps, and install new refuse-burning equipment in city housing projects. But all this costs millions of dollars and New York City, like most cities in the United States, is already in a difficult financial position. For that reason—finances—governmental units are often the worst polluters. Moreover it is socially and politically more acceptable for critics to focus their wrath on industrial groups. Certainly this is often justified, as is shown in the incredible story of Union Carbide's plant in Marietta, Ohio, which for over six years resisted informal and ultimately formal orders by the state and federal government to reduce its level of emissions. Only when the serious possibility arose that federal authorities would seek to shut down the plant did Union Carbide take effective action.

Another problem concerns the inability of the federal government to deal effectively with other governmental agencies including some within the federal government itself. The threat of closing down a publicly operated facility is not nearly as effective as a similar threat made to a privately operated corporation. How can you tell people in a town to stop flushing their toilets because the local sewage treatment plant is not functioning as it should? Sometimes a ban is placed on the issuance of new construction permits and sometimes government officials in the federal government will threaten vague forms of punishment or at least seek to embarrass city or state officials in order to compel compliance. That is what happened in early 1971 when the Director of the Environmental Protection Agency criticized Atlanta, Cleveland, and Detroit for their failure to comply with water quality standards set by the federal government. However, as officials in the criticized cities were quick to point out, their efforts had been hampered by broken promises of support from the federal government. Who then can punish the federal government for its failure to follow through on its commitments? Who can take action against federal installations such as navy and army bases, or perhaps the biggest public or private offender of all, the US Army Corps of Engineers, with its massive dam and canal projects that often lead to serious disruption of the environment? Even when state and local governmental agencies desire to clean up their own messes, they usually lack the financial means. The same often holds true, though to a lesser extent, for industrial polluters. As a result, even if stringent laws were passed, there is no guarantee that such legislation would remedy the problem because of the financial burdens such cleanup entails. Therefore effective pollution control will necessitate subsidization as well as direct legislation.

Subsidization

Because of the financial plight of our cities and the reluctance of many firms to spend large sums of money on projects essentially unproductive in terms of private profits, some kind of subsidy may be necessary to implement antipollution laws. For the most part, the bulk of this financial help will have to come from the federal government, the only governmental unit with access to the large sums involved. Such aid could take the form of tax credits. In some cases the polluter could be provided with supplemental funds that would otherwise not be available to him. Hopefully such support will raise the likelihood of compliance.

There are almost as many different kinds of tax credit available for industrial and communal polluters as there are taxing authorities.[20] Typically the municipality or state guarantees the polluter that the assessed

---

[20]*Industrial Incentives for Water Pollution Abatement* (New York: Institute of Public Administration, February 1965), chaps. VI–VII.

value of his real estate will not be increased if pollution abatement equipment is installed. Some tax jurisdictions go further and permit a deduction from the tax base if such equipment is purchased. Sometimes such a deduction is made on real estate taxes, but more frequently it is applied to state or city income taxes. Such deductions are usually embodied in the form of accelerated depreciation or a more rapid write-off of the cost of industrial equipment from profits. The concession may also take the form of an investment tax credit. This is similar to the credit that was permitted on all federal income taxes for new manufacturing equipment purchased from 1962 to 1966. The investment tax credit itself was abolished in late 1966, but Congress agreed to continue a special investment credit for just pollution control equipment. Under such a system, manufacturers are permitted to deduct a sum equal to a stated percentage of the cost of the new equipment from their taxes. Such a concession is in addition to the regular amount deducted as depreciation. In this way investment credits reduce the size of the tax payment sent to the government.

Some states have also considered the use of investment credits. Unfortunately at the state level the net effect of such a concession is usually quite low because state and local income taxes are generally not high. Moreover, money paid as state income tax is considered to be a deductible expense from corporation profits subject to the federal tax. Therefore corporations end up with an even smaller portion of the rebate they receive on their state taxes since they would normally deduct close to 50 percent of that state tax from the money they owe on their federal tax. The net impact of state concessions is about one-half of what it appears. Thus if the state tax rate is 5 percent, the credit 5 percent, the cost of the machine $100,000, and the firm's profit $500,000, this will mean a net saving to the firm of only about $1,250, not $5,000.

Federal government subsidies take the form of loans, guarantees, and grants. The federal government makes available a variety of low interest loans for the purchase of pollution equipment. Occasionally it may do no more than guarantee a loan, but this is still a form of subsidy. In some cases outright grants are made. Such grants in economically depressed areas may constitute up to 80 percent of the cost of new treatment facilities. Here the government's purpose is to provide jobs and economic security as well as to eliminate pollution. The federal government is also experimenting with what are called "Demonstration Projects." Washington will grant up to 55 percent of the cost of a water cleanup project if the program is related to a basin-wide effort and includes financial participation by the state. If there is no coordinated program, the federal government will provide only 30 percent of the cost. The average size of most grants is about 40 percent.

There is considerable pressure to increase the federal share of construction costs for pollution control even further to 90 percent. This is the ratio used to finance the federal highway program. However, anticipation of this potentially generous federal participation has created serious problems. Some states report that they are having trouble obtaining new appropriations since fiscal officials argue that it will cost the state less if they wait only a few more years. Therefore, the states that are eager to act now want a promise that the federal government will compensate them retroactively. Federal authorities are opposed to any such promise. How far back does this mean? Moreover if the federal government has to provide money for projects that are already completed, it has that much less money to stimulate new construction.

For many municipalities and states, however, the question of whether the federal subsidy should be 30, 55 or 90 percent of total construction costs is treated as a secondary issue. The primary issue is whether the federal government will provide anything at all. Under the pressure of Congress, especially from anti-pollution buffs like Senator Edmund Muskie (see Part 5), at least there has been an acknowledgment that large sums are required for pollution control. Consequently, beginning in 1967 Congress authorized the expenditure of increasingly larger sums for water and air quality programs (see Table III). Unfortunately, as the war in Indochina claimed larger and larger sums of money, Presidents Johnson and Nixon found it necessary to hold back budget expenditures in other areas, pollution control being prime among them. Invariably Congress would point to the ambitious authorization. The president would respond with a budget request that occasionally was one-fifth the size. Congress would counter with a higher proposal but, as seen in Table III, the president's figure usually prevailed. That was the pattern at least until fiscal 1970 when Congress threatened to appropriate the full amount originally authorized. A compromise was finally worked out in which Congress insisted on an appropriation of $800 million, or about four times what President Nixon wanted.

Eventually, in 1970 President Nixon himself came to embrace the need for greater expenditures in what was announced as a $10 billion program. Starting slowly because the expenditures for Vietnam had not yet diminished significantly, he proposed the revised schedule of federal outlays shown in Table III. As his critics were quick to point out, the $10 billion program would in fact involve only $4 billion of federal money. The remaining $6 billion would have to be provided by the states and cities according to Nixon's proposed 40–60 percent formula. Moreover the largest sum in Nixon's program for any one year was to have been $800 million in 1975. To say the least, President Nixon's plan compared unfavorably with the $1 billion originally authorized by President

Table III: *Federal Authorization and Appropriation for Water-Pollution Control (in millions of dollars)*

Fiscal Year	Author-ization	Presidential Request	Appro-priation	President Nixon's 1970 Request	President Nixon's 1971 Request
1968	$ 450	200	203		
1969	700	203	214		
1970	1,000	214	800		
1971	1,250		1,000	40	
1972	1,250			200	2,000
1973				440	2,000
1974				760	2,000
1975				800	
1976				760	
1977				600	
1978				360	
1979				40	

Johnson and the Congress for 1970 and the $1.25 billion authorized for fiscal 1971. President Nixon subsequently increased his requests to ask for a total of $2 billion a year for 1972 to 1974 and thereby accepted the Senate's pressure for more money than the skimpy $40 million he had originally requested.

A similar reluctance has been shown by the executive branch in meeting congressional requests for funds for air-pollution control, although the sums involved have traditionally been much smaller. So for fiscal 1970 Nixon asked for $106 million, but the Senate wanted him to spend $275 million. The same difference in approach exists in regard to solid waste control.

The difficulty in obtaining adequate funds for pollution control has complicated the efforts of the cities and states. It is bad enough when inadequate sums of money are promised in the first place, but when promises are made and then broken so that actual appropriations turn out to be lower than authorizations, there may be complete demoralization. As we have noted, this is at least a partial explanation of why Atlanta, Cleveland, and Detroit are lagging behind in their efforts to reduce water pollution. Agreement between the three cities and the federal government was only reached in June 1971 when the federal government finally agreed to come through with a grant of about $450 million. By the end of 1969 because the federal government had failed to provide the money it had promised, various cities and states had already been forced to find alternative financing of over $800 million for water-pollution control expenditures which were supposed to have come from the federal government. The federal government has promised that eventually it will compensate the cities and states for this money out of future appropriations. In turn this has the effect of reducing the money available for new projects. Yet the needs increase and because the costs involved are so high, subsidies from the federal government and the states are often necessary to make possible any local or private construction of pollution control facilities. It is easy to envision the panic in a one-industry town when the factory manager declares that any effort to make the factory curb its pollution will force it to leave the area. This may be a bluff, especially as pollution standards are raised throughout the country or as federal controls are instituted. Still, the availability of financial aid in the form of low-interest loans or outright grants is likely to reduce the tension on the bargainers and might prevent the closing of plants such as the Olin chemical complex in Saltville, Virginia, the Kuhlman Corporation's processing division in Detroit, and the Automation Industries' milling facility in Gardena, California. These companies all found that remedying their pollution emissions was too expensive to permit them to continue operations. Proposals have been made to permit federal subsidies for joint municipal–industrial treatment that might help those plants with water treatment problems. This would avoid direct grants to private industry but at the same time would occasionally make possible economies of scale and of neutralization. Economies of scale often result when one large plant is built, rather than two small ones, because then it is possible to operate with one pump instead of two, one manager instead of two, and so forth. Thus there is often a saving in both construction and operating costs. Neutralization occurs when the right mix of acid wastes happens to combine with the right mix of basic wastes, resulting in a harmless product which may reduce the need for additional treatment. Thus when a plant of the Allied Chemical Corporation near Syracuse, New York, was closed down because of a strike, Onondaga Lake developed a bad odor. When the plant could no longer pump its chlorine, calcium chloride, and lime water wastes into the lake, there was nothing to offset the waste from the nearby municipal sewage plant. However, when the strike was over and both plants were again working in tandem, there was no serious problem.

England and Japan encourage joint usage provided the industrial wastes do not adversely affect the sewers or alter the normal sewage treatment process.[21] Subsidies, however, may not be warmly regarded

[21]Mrs. Lena Jeger, Ministry of Housing and Local Government, *Taken for Granted; Report of the Working Party on Sewage Disposal* (London: Her Majesty's Stationery Office, 1970), p. 44.

by other taxpayers, who may resent the diversion of their funds to the private use of industrialists.

Edwin Mills, formerly of the President's Council of Economic Advisers and now at Princeton University, argues that between a tax credit for investment in pollution control and a subsidy for the same thing, he would prefer the tax credit (see Part II). In his view, a subsidy for investment tends to discourage the use of other pollution abatement techniques. He suggests for example that in some cases instead of a capital expenditure for some device, air pollution might be better controlled by switching to nonpolluting fuels, that is, from coal to gas. Other authorities fear that the use of subsidies to prevent pollution could lead to a kind of gamesmanship on the part of polluters. By threatening to open up new and dirty blast furnaces, some industries may be able to coerce payments for projects they did not really intend to build. This would be like the farmers who earn their income by agreeing with the Department of Agriculture *not* to plant corn.

Incentive Pricing

A third and somewhat novel solution is to utilize a system of economic incentives so that the polluter is forced to come to terms with the social costs of his pollution. This is called "internalization"—the creator of environmental disruption is made to take upon himself (or internalize) the costs of his actions which previously he had passed on to society at large. Some economists have argued that only when the private costs of the polluter are also adjusted to include the social damage caused by his pollution will he willingly do his best to clean up his effluent (see Part III). Depending on the nature of his pollution and the possibility of joint action with several polluters, the polluter can decide himself without the need of a government edict whether or not it is better (or cheaper) to clean up his own pollution or join in some cooperative effort. There are formidable technical difficulties involved in implementing such a proposal, but it has considerable merit, at least as a supplemental measure along with legal regulation and government financial aid.

In discussing the market and price system we stressed that it is incapable of preventing pollution as long as it lacks certain basic information. To function properly, the system must be adjusted to take account of external diseconomies, differences between social and private costs and under- or overvaluation of inputs and outputs. These shortcomings must be eliminated. This means supplementing the operating costs of municipal and private enterprises with the costs arising from any pollution they create. This in turn means obtaining accurate estimates of the social costs of various types of pollution. A system has to be developed that will make it possible to trace the origin and extent of various effluents. Only if this is done will it be possible to incorporate the costs of pollution in with the other clearly specified operating costs of the polluter.

Assuming that some group—federal, interstate, regional, cooperative or private—can make such assessments, the cause of pollution abatement, theoretically, will be far advanced. The assessing group must be able to calculate the damage created by a given polluter and a charge be made in proportion to the damage created by the resulting air and water effluent. Once that information is cranked into the cost calculations of the particular producer, the system may be largely self-sufficient and self-correcting. As is true with most operations in the market system, the polluter will then have before him a tabulation of all his costs. If it is determined that his pollution has had no negative effect on the rest of society, the producer can be allowed to operate as before. If damage has clearly been done, then the polluter should be made to pay the extra costs for pollution control. Once he knows the value of his damage, he has several choices: 1. He can cut back production and thereby eliminate the pollution problem. 2. He can install control equipment himself—profitably—if the operating cost of the equipment plus amortization is less than the penalty he would have to pay for his pollution. 3. If it is cheaper for him to pay the penalty than cut back production or reduce his pollution, then he simply would pay the damage fees to an outside body which would in this way acquire income to finance its own pollution control. Presumably this latter situation would exist when there are economies of scale in pollution control that make it more efficient for one treatment plant to handle the pollution of several factories than have each plant treat its own wastes.

Doubtlessly there will be shortcomings with such a scheme, just as there are in the normal market system itself. Everyone acknowledges that the market system needs help from the government to contend with monopoly and other imperfections. Most likely, the government would have to step in to supplement pollution abatement efforts. Some specific restrictions or subsidies will certainly be needed at times to augment the purely economic incentives. Even the harsher critics of the market system usually acknowledge that regulation solely by means of government edict is not a better solution. In fact most socialist theoreticians advocate that a market system be utilized in a socialistic society to make the millions of decisions which must be taken every day.[22] That operation of a state-owned economy is not possible without the use of traditional economic tools is indicated by recent economic reforms in the U.S.S.R. Nor as already indicated is there anything to suggest that Soviet edicts have been

[22]Oskar Lange and Fred Taylor, *On the Economic Theory of Socialism* (Minneapolis: University of Minnesota Press, 1938) p. 8.

any more successful in preventing the pollution of the air and water. Because a person, a city, or a manufacturer may cause pollution in so many ways, frequently unintentionally, a police network without self-enforcing mechanisms of the type just outlined would be too immense, too impractical, and too arbitrary.

Assuming that costs and responsibility can be properly assessed, there is much that such a system can do—some of which today depends more on public spirit than on economic necessity. For example we have seen how air control authorities have become concerned over the increase in the sulfur dioxide released into the air. Some managers of electric utilities as good citizens have switched from coal to oil to gas, both of the latter being lower in sulfur than coal. Others have been forced to make the change by law. Yet if, as President Nixon proposed in 1971, emitters of sulfur dioxide were obligated to pay a fee for every ton of sulfur dioxide they release into the air, the decision to switch to gas would be taken much more readily. But since it is entirely possible that the use of gas in all circumstances and at all times might not be the most efficient manner of handling the problem, an alternative decision could be made by the polluter without seeking government approval. Thus an electric utility located in a coal mining district where coal is cheap might find it more economical in the long run to continue the use of coal, but to collect the sulfur from the stacks and sell it as a by-product. Similarly if economic penalties were imposed on the use of sulfuric acid for cleaning steel plate, more steel companies would probably find it expedient to use hydrochloric acid instead. Republic Steel and Great Lakes Steel Companies have discovered that hydrochloric acid can be easily reconstituted after it has been used. Moreover its use avoids the necessity of having to treat the difficult-to-handle sulfate wastes produced by the former acid.

Because of the complexity of economic interrelationships, such arrangements sometimes may be easier to propose than to implement. Thus the electric utility will be stimulated to extract sulfur for resale if the price of sulfur is high enough. If, however, steel manufacturers are discouraged from using sulfuric acid, the price of sulfur will probably fall. In sum, economic solutions, just like political solutions, may not always be as simple as they first appear.

It is necessary to remember that polluters respond readily to economic incentives in ways that might not always be anticipated at the time by government administrators. Part of the task of the government is to anticipate what these reactions might be and utilize them so that pollution is reduced. Sometimes this is the accidental result of an innovation that originally had nothing to do with pollution control. Nevertheless when profit-seeking businessmen utilize this innovation, they reduce environmental disruption along the way. For example, the problem of unked cars is likely to become somewhat less bothersome in the future because Pittsburgh Steel Company has discovered a way to increase the use of scrap metal in the oxygen steelmaking process. The main stimulus for such innovation was that scrap had become so cheap. In 1966 it was almost half the price of what it sold for before the introduction of the oxygen system.

The cheap price of scrap actually stimulated the widespread construction of electric steel furnaces, which utilize large quantities of scrap. By 1970 the demand for scrap from the electric steel furnaces had pushed the price of scrap up so high that some of the new furnaces were no longer profitable. Simultaneously this higher price had a positive effect on the scrap dealer. Once more it became attractive to buy scrap to sell to steel mills for recycling. Some junk dealers purchased portable flateners. This equipment crushes cars and enables dealers to move them, compactly and inexpensively, to urban stripping centers where they are passed through shredder machines that chop automobiles into small pieces. A magnet is passed over the remains to remove the iron and steel fragments, which are then bundled off to the steel mill relatively free of noniron particles. A similar device, designed by a Belgian firm, uses the principle of cryogenics, or deep cold, to remove scrap iron and steel. Since metals of different densities react differently to cold, liquid nitrogen is passed over a bundle of metal, freezing it to as low as $-310°$ F. At such temperatures ferrous metals become brittle and splinter like glass when put in a shredder. Nonferrous metals, which must be removed from the scrap bundle if the steel is to be reused in the steelmaking process, are more elastic under such cold and therefore remain intact, so that a magnet can remove the ferrous from the nonferrous metals more efficiently under this system.

While some of these innovations may be fortuitous, it is clear that the government can do much to stimulate other methods of pollution control through use of subsidy and other economic incentives. For example, to induce even greater use of scrap, the government or another pollution-control body could subsidize the processing of such scrap out of automobile excise taxes. Another alternative is to include a $100 deposit charge in the price of every new car. In exchange, the new owner would be given a certificate of deposit which would be passed on to each subsequent owner. Upon turning his car over to the junkman, the final owner would be able to reclaim his deposit by showing state officials that his certificate had been suitably stamped by the junkman. Such a procedure was proposed to the New York Legislature by New York City officials in 1971. This system and others now being introduced experimentally by

General Motors could help restore the self-perpetuating cycle of scrap to steel to automobile to scrap that existed before, to hold down the growth of automobile graveyards.[23]

Utilizing somewhat similar principles, the meatpacking companies around Omaha managed to convert "an industry and community liability into a revenue generating resource" that also solved a serious water pollution hazard.[24] Previously almost all Omaha's meatpacking plants discharged their waste untreated into the water. The municipal system was bypassed because it could not handle such large volumes of animal waste. A solution was designed whereby a separate treatment plant was constructed to handle the meat plant waste and send the processed effluent on to the city sewage treatment plant where it could be further purified. A nonprofit corporation was formed to float bonds in order to finance construction. It is estimated that operating costs will be more than met by the packers who will pay for the use of the facilities and by the resale of recovered waste products, so that extra cost to the consumer will be minimal.

Suggestions have also been made periodically that the process of incineration be combined with electrical generation. Certainly it seems wasteful to burn coal or oil to generate heat that is then sent up the smoke stacks of incinerators. So far such combined operations seem to have been more successful in Europe, although Atlanta, Georgia and Norfolk, Virginia, have made limited use of such methods. A similar combination of processes could be designed to reduce thermal pollution. In some cities, as a by-product of the generation of electricity, electric utilities also supply steam for heating and air conditioning to buildings in the downtown area. It might also make sense for utilities at the same time to supply hot water for tap use. All that would be required is that a second pipe be added alongside the existing water supply pipe. A simpler arrangement could probably be designed where a steam pipe already exists for centrally provided heating. It is a waste to dump cooling water, often uncontaminated from power plants into water courses while other energy forms are consumed to warm up cold water. This system should be particularly effective in areas of population concentration like New York City, Chicago, or Miami Beach.

Many other methods have been proposed to stimulate pollution control. When California decided to require that all new cars be

[23]*New York Times*, October 12, 1970, p. 29.
[24]Federal Water Pollution Control Administration, U.S. Department of the Interior, *The Cost of Clean Water; Economic Impact on Affected Units of Government* (Washington, D.C.: Government Printing Office, January 10, 1968), pp. 116–19.

equipped with pollution abatement devices, it created a tempting economic incentive for accessory manufacturers. For a time the big three automobile producers argued that they could not develop such devices for their new cars. However when the state declared that it would proceed with implementation of the new car law anyway, Ford, Chrysler and General Motors found they could create a device after all. It was either make their own or purchase equipment from outside suppliers. The threat of even more stringent laws or taxation on gasoline-powered cars has also sparked renewed interest in the electric-powered automobile. This in turn has provoked the oil industry into seeking more effective methods of eliminating pollutants from automobile exhaust.

Undoubtedly similar results would also take place if the right combination of economic and government pressure were applied to other problems. Sometimes just the threat of ultimate government action is enough to induce some pollution-control device that on occasion may even turn out to be self-financing. For example, as pressure increased for the owners of fuel tankers to cease the flushing of oil tanks, Cities Service Company decided to do something about it. Cities Service came up with a design in which tankers burn their cargo residue as fuel in their own diesel engines. This eliminates a major cause of harbor and waterway pollution besides producing a fuel saving. Similarly more can be done through government encouragement to stimulate productive use of the sludge from sanitary treatment plants. Sludge is high in fertilizing capacity. Some cities such as Las Vegas and Amarillo now sell it for use on golf courses and parks. There is no reason why sludge could not be used elsewhere as long as sanitary conditions are observed. Similar reuse of processed sewage takes place at University Park, Pennsylvania, where the effluent is sprayed over a pine forest and a golf course. This returns nutrients to the soil and filtered water to the water table. Milwaukee has been packing and selling its sludge around the country as a fertilizer called Milorganite. Even the Russians use sludge as fertilizer for certain agricultural products.[25] Lubbock, Texas, is constructing a system of man-made lakes that will be filled with reclaimed waste water. From the lakes, the water will then be taken to irrigate cotton fields, after which it will percolate through rock and sand. It is hoped that this will serve as tertiary treatment so that water pumped from the recharged groundwater can then be added to the city's drinking water.[26] For a time, St. Petersburg, Florida, paid the International Disposal Corporation, a private firm, $3.24 a ton to process its waste. This was half the cost of conven-

[25]*Rabochaia Gazeta,* July 3, 1970, p. 3; *Priroda,* August 1964, p. 73; Marshall I. Goldman, *Environmental Disruption in the USSR* (forthcoming, 1972), chap. 3.

[26]*Water Pollution Control Federation Highlights,* October 1970.

tional garbage disposal. In a plant designed by Westinghouse Electric Corporation, the waste from St. Petersburg was then processed and sold as compost to nurseries, home gardeners, and truck farms. Unfortunately there were serious odor and mechanical difficulties which eventually forced the operation's termination. However in 1970, Chicago worked out a less complicated program whereby its sludge was sent 160 miles downstate to Arcola, Illinois. Each day a thirty-six-car train carried 1,500 tons of Chicago's treated sanitary waste to three farmers in the area who applied it to alfalfa, soybeans, and corn on their land. Ultimately over 1,600 acres will be fertilized in this way and the project will consume about 20 percent of Chicago's sludge. Disposing of it on farmland costs about $15 a ton, whereas alternative methods would cost as much as $50–$57 a ton.[27] The trick is to find someone who is willing to accept someone else's sewage and who also has neighbors who are away from home most of the time!

Unfortunately there are also innovations in the opposite direction. We have already noted the growing use of nonreturnable bottles and cans. This may be more convenient for the individual but it results in litter along our beaches and highways. To reverse this trend Coca Cola has raised the deposit it charges from 2¢ to 5¢ per bottle. The hope is that the additional pennies will prove enough of a stimulus to make people return the bottles. Other bottlers have embarked on similar programs. In many cities, even canners are offering ½¢ for every returned aluminum can. If such systems do not work, it may ultimately be necessary to tax the sale of such items or require deposits on all soft drink or beer containers, as Oregon plans to do after October 1, 1972, in order to induce the utilization of returnable containers again.

The proper use of economic incentives can also do much to bring about a more rational use of natural resources. By recognizing the scarcity value of water and by making consumers pay full value for it without subsidies from property taxes or electric power revenues, it should be possible to reduce the use of water for irrigation purposes. In response to pressure from the agricultural bloc, over the years government policy has provided cheap water for agriculture. This has resulted in a topsy-turvy situation where in the dry lands of Arizona it is possible to buy water cheaper than in St. Petersburg, Florida.[28] This is done despite the fact that the deep wells that supply Tucson are drying up. It is also ironic that

[27]Jerome Goldstein, "Transporting Wastes to Build Soils," *Compost Science,* September–October 1970, p. 22; Warren J. Papin, "Soil Enrichment Express," *Water Spectrum,* Winter 1970, p. 16.

[28]William Bowen, "Water Shortage is a Frame of Mind," *Fortune,* April 1965, p. 146.

water used for irrigation evaporates at a much faster rate than water used for industrial purposes. Therefore agricultural water cannot be reused or reprocessed as much as industrial water, much of which is simply used for cooling purposes. Moreover, water used in irrigation adds less to the gross national product than water used in industry. Some estimates indicate that water used in agriculture adds fifty dollars to the GNP per acre-foot of water used, compared to three thousand dollars for water used industrially. Water is so cheap that it usually does not pay to cement or cover irrigation ditches. Inevitably this only increases the evaporation rate. Obviously the introduction of a higher price for irrigation water would induce curtailment of water used for such purposes and divert it to uses which contribute more to the country's well-being.

Similarly, existing economic incentives favor the consumption of new raw material resources rather than the recycling of already used products. Ecologists and economists have long warned that there is no such thing as waste disposal. Everything goes back to the air, water, or land in some form or other. Rather than discard our resources in some unusable form, it would be much better to reprocess our effluent so that it can be recycled. Presently, however, many manufacturers are discouraged from doing this by the existence of laws and economic incentives applied without regard to their ecological consequences. For example, depletion allowances provide miners and oil drillers with tax rebates when they dig new ore or drill new wells. From an ecological point of view, instead of rebates there should be added taxes for the use of virgin materials. If any rebates are to be given, they should be given, for example, for the reuse of scrap or oil used in automobiles.

It is not just the tax laws that frustrate the recycling process. At one time it was common for garage and gasoline station owners to collect and sell the used oil drained from automobile crankcases during an oil change. Such oil had a value because it could be filtered, reconstituted, and sold for reuse at a profit. However new labeling restrictions made the oil more difficult to sell (who wants to put "used" oil in *his* car) at a time when the costs of rerefining were rising. This led to the closing of most rerefining operations across the country. Those that remained found it necessary to charge about 3¢ a gallon to haul the oil away. When the gasoline dealers and garage owners found they would have *to pay* to have their wastes removed instead of *being paid* for it, most of them did what most of us would have done under the circumstances—they dumped it down the drain. Unhappily few sewage treatment facilities are equipped to process oil, and as a result there has been a serious increase in the discharge of oil in the vicinity of many sewage treatment plants. Much can still be done to reverse such irrational practices and stimulate waste recovery programs. Instead of glorifying and subsidizing the Marlboro Man, perhaps we should glorify and subsidize the junkman.

Economic inducements can also be adopted to stimulate a more rational disposal program for water and air wastes. For example, natural and artificial sanitary treatment facilities are generally underutilized during the winter months and during the early morning hours between 2 and 6 a.m. The water is usually colder at such periods and less waste is thrown into the system. Therefore the oxygen-carrying capacity of the water is greater. In addition the winter water level is usually higher than it is in the summer when there is less precipitation. Therefore it would be economically expedient to reduce sewage charges during these periods so that manufacturers will be stimulated to hold back some of their sewage until the water course is better able to absorb it. The savings earned from the reduced sewage charges can be applied to the construction of storage lagoons to hold the effluent. Similarly it may be possible to tax automotive traffic in the morning in order to shift commuters to subsidized public transportation. This could be done by increasing tolls on bridges and highways during rush hours. Since temperature inversions are most likely to occur in the early morning hours, this is precisely when automotive fumes should be reduced.

Undoubtedly there are other presently wasted materials that could be put to productive use given the proper stimulus. If economic conditions warranted, presumably the United States Steel Company in Gary, Indiana, would make even more of an effort to reclaim the approximately 13,750 pounds of ammonia nitrogen, 1,500 pounds of phenols, 1,700 pounds of cyanide, and 54,000 pounds of oil that it has been sending out each day into Lake Michigan via the Calumet River. As it is, extreme pressure has often had to be applied to U.S. Steel to produce action—at the older South Works plant in Chicago, the Illinois Attorney-General found it necessary to institute a court suit to force U.S. Steel to recycle its water wastes. Under the plan, which will be particularly expensive because the plant is so old, at least $12 million will have to be spent in an effort to eliminate completely the discharge of cyanides, phenols and ammonia wastes into Lake Michigan.

That steel mills can also be made to clean up their wastes through economic stimuli as well as through legal pressure is not an idle dream. The best evidence of this is the successful utilization of economic incentives and government cooperation to clean up the waters of the Ruhr Valley in Germany (see Part III). Here, in fact, is the perfect example of a pollution control system using the best of all techniques. Under the guidance of unique cooperative associations called *Genossenschaften,* industries and municipalities have combined to coordinate and control their water wastes. Through the use of proportional sewage and water charges, the polluters are induced to decide for themselves when it is better to pay a sewage charge and when it is better to process the sewage themselves. It has been discovered that in many instances it is better to pay a sewage charge because economies of scale make it possible for the *Genossenschaften* to process the wastes of several polluters more efficiently than if each tried separately. The factory charges are then used to finance the cost of operating the cooperative facilities. In some cases through adept plant location, regional planners may be able to alleviate the pollution problem without the need for expensive treatment. It sometimes happens that the wastes of one factory neutralize the wastes of another polluter.

The success of the Ruhr system is indicated by the fact that although the annual low flow of the Ruhr and its tributaries is not much more than one-half the low flow of the Potomac River, the Ruhr River basin is now able to handle the pollution from 40 percent of all West German industry with ease. Steel industries in the area use their own water over and over again. Thus as Allen Kneese points out in Part III, it takes only 2.6 cubic yards of water to make one ton of steel compared to the 130 cubic yards it takes elsewhere. The end result is a self-supporting system that operates with a minimum of government interference and a maximum of efficiency.

Despite what might be called the beauty of the *Genossenschaften* system, it would be wrong to assume that such systems can be instantaneously adopted elsewhere. For example, the Ruhr itself has not been especially successful with air pollution control. The state of Vermont has decided that after July 1972 no factory may discharge into a watercourse without a permit from the state's Water Resources Department. A fee will be charged for the issuance of this permit, and the amount of the fee will be related to the waste load created by the discharge. This could lead to a system of effluent charges which ultimately duplicates the Ruhr system. Something similar has been introduced by the Metropolitan Sanitary District of Chicago. Nonetheless, efforts to introduce the *Genossenschaften* system on a large scale in the United States will probably be resisted because the main challenge in trying to implement such a system in the United States is more political than economic.

The first hurdle that must be overcome is the winning of political support from the numerous skeptics who doubt that economic controls are workable. They point out that many polluters distrust the use of economic controls—some distrust them because they do not understand them.

Others cite the fact that economic controls have not always worked well. Occasionally arbitrarily applied taxes and subsidies have solved one set of problems only to create a whole new set of distortions. Thus some critics fear that the use of pollution charges will bring about just the opposite of what is intended. They will make it easier for organizations to pollute with a clear conscience. In calling the economic charges "a license to pollute," these critics fear that the factory will be able to pollute, pay its fines, and then claim that it has fulfilled its civic responsibility. Such critics should recognize, however, that when they urge a complete ban on discharge into water or air, they are applying a value to air and water in exactly the same way as do the economists they criticize. Without knowing it, they have simply said that in their scheme, air and water have an infinite or at least an extraordinarily high value. Such a valuation may be the correct one under a particular set of circumstances, but those who implicitly attach such values should realize what they are doing.

An even more serious difficulty has to do with the problem of political jurisdiction. Although we normally think of our political system as quite flexible, our cities and states as already mentioned are not usually structured around complete water and air basins. Therefore any control organization must make provision for the overlapping of political boundaries and the transfer of adequate authority and funds to new and specially created units. As our problems become more regional in nature, there is also the danger of excessive duplication and contradiction. Lately there has been a tendency for such agencies to proliferate since a regional agency that may be large enough to cope with one problem may be too small to handle another. Eventually there may well be a reluctance to disburse power to such groups for fear that they and not the usual organizations will claim all the power and money.

Not only must there be a skillful use of political talent in order to obtain political cooperation and fiscal subsidization, but also it may be necessary to restrain some of the overeager economists. Occasionally some economists do go to extremes. There is a danger that some economists might view such experiments as an opportunity to apply classroom theory to the external world in unrestrained form. In fact, the *Genossenschaften* in the Ruhr have been quite restrained in the use of economic incentives. They have adopted the underlying economic principles while simultaneously resisting the temptation to become unduly involved in the theoretical minutiae of the problem. They have managed to find the optimum combination of theory and pragmatism, with considerable emphasis on the latter. No effort is made to determine precisely how much pollution is created by a given firm at a given moment. They have rec-

)gnized that engineering technology has not yet devised a cheap means of providing instantaneous and accurate data about pollution emissions. Therefore it is not possible to vary effluent charges precisely in correspondence with the social cost involved. If the *Genossenschaften* did make an attempt to provide such coverage, there is the real danger that the cost of administering such a system would be greater than any possible savings that might result. For some time to come, it may be wiser to use a less elegant system but one that accomplishes the basic purpose of air and water pollution restraint.

Because the problems and complexities of pollution are so enormous, it is reassuring to know that there are places like the Ruhr Valley, where some progress has been made in reducing or at least coping with pollution. In addition it is encouraging to note that with a mixed package of political and economic incentives and penalties, English authorities have managed to improve conditions in the Thames enough so that by the late 1960s fish returned, after an absence of almost a century. Similarly air pollution has been reduced. In response to the crippling smog of 1952, Parliament passed the Clean Air Acts of 1956 and 1968, which regulated the air emissions of industry and the burning of soft coal in home fireplaces in certain key areas of the country—as a result, there has been no killing smog for the last decade or so. London now has 50 percent more sunshine in winter than it had in the mid 1950s, and seventy additional bird species have been sighted in recent years. If London, one of the most heavily populated and industrialized cities in the world, can make this much progress, certainly there is hope for other areas of the world. Some other instances citing progress in the battle against pollution are presented in Part III.

SOME PHILOSOPHICAL COMPLICATIONS

In our efforts to eliminate or reduce the environmental disruption that most concerns us, we must remember that one man's garbage may be another man's bread. This holds true on the international as well as the domestic scene. For example, in the late 1960s when India was faced with a serious grain shortage, there was no question about the urgent need for the use of DDT—the choice for India was between starving human beings and dead or deformed wildlife. A similar tradeoff faced the Indians in their fight against malaria, until recently one of the most crippling diseases in tropical countries. DDT has sharply reduced if not curbed its incidence. When malaria and other pest-engendered diseases give rise to social costs that equal or exceed those of environmental disruption, it is not surprising that the lesser of two evils may be chosen. If in the process of eliminating malaria and pest problems there is a coincident increase in DDT in pelicans and human beings, this is regarded as a relatively small price to pay. The determination of many advanced countries to curb the production of DDT is viewed with astonishment in countries like India. To the Indians, a person with a high concentration of DDT in his system is preferable to one who is no longer alive because he died of malaria. As one Indian put it, not to use massive quantities of DDT to kill malarial mosquitos would be a form of genocide.[29]

Unfortunately the consequences of India's use of DDT are not limited to India. Birds that have absorbed large quantities of DDT in India may migrate to other areas of the world and enter into the food chains. Thus one country's action in regard to environmental disruption may easily have international implications. The decision of the Russians and French and English to continue with their SSTs is something that affects us all because of the plane's impact on the stratosphere and should be protested by the world community. In this case the whole world suffers when one or two countries insist on their pursuit of "technological progress." In other instances the issues are not so sharply defined. Thus it should be remembered that while we may fret over India's continued use of DDT, the Indians are equally concerned that our banning of DDT production will have adverse effects for them.

While the contrast is not nearly as sharp, there are also occasional differences in attitude toward the question of environmental disruption within our own country. It is sometimes said both correctly and incorrectly that pollution is primarily a middle and upper class concern. Some even argue that it is a "cop-out" used to avoid facing issues like poverty and racism. Unfortunately there is often an element of truth to such charges, but a black child is just as vulnerable to polluted air and water as a white child—probably more so because pollution is usually more intensive in center cities.

In connection with such charges it is necessary to look briefly at the question of who will bear the burden of clearing up environmental disruption. It is a question that is impossible to answer precisely, but as we shall see, there is good reason to believe that the poor will bear more than their share of the burden in the clearing up of certain types of pollution, especially if some of the economic incentives proposed above are adopted and products are made to reflect not only the private but also the social cost of their production. Ultimately most of the burden will fall on

[29]*New York Times*, February 15, 1970, Sec. 4, p. 4.

the consumer in the following way: When the market situation is competitive, the bulk of the costs will fall on the producers themselves. Their profits will drop and this in turn should eventually discourage the flow of new capital and the expansion of productive capacity. To the extent that this results in a fall in output or the closing of some businesses, prices of the affected goods should rise, transferring at least part of the cost to the consumer.

In industries and activities where there is less competition or where the demand curve is inelastic, the cost will be transferred directly to the consumer. Conceivably, this will lead to a search for product alternatives. This will not be easy since we assumed initially that the demand for the original product was inelastic, which means there are no readily apparent substitutes. But if the costs of these inelastic goods rise and the quantity of the goods purchased remains much the same, the disposable income available for the purchase of other goods will be reduced.

Whether the producer bears the cost of pollution control or passes it on, the ultimate result is that the prices of goods to the consumer will rise. Unless total personal incomes increase, this will lead to a fall in real personal consumption. In effect, it will be as if a new sales tax were imposed on all consumption. Just like war expenditures or new highway construction, such funds will go for a type of nonconsumption good—pollution control.

Like any sales tax, this will probably mean a proportional fall in actual personal consumption for all income groups. But since lower income groups spend a larger proportion of their incomes on goods than upper income groups and a program of pollution control will probably be reflected primarily in higher prices on consumer goods, the cost will fall disproportionately on lower income groups. This is what we expect with any regressive tax system. To the extent that federal funds are used to defray municipal sewage costs, some of the revenue for the necessary facilities will be obtained through the progressive income tax. This would probably place more of the burden (i.e., reduced consumption) on the upper income groups. Those who believe in progressive taxation would approve of such an outcome since, as we saw from Table II, improving pollution control in various government sectors will necessitate larger expenditures than those needed in industry. However, if we are to resort to a user's charge, which seems to be the best way of curbing our use of water and air, the progressive income tax can only be used to a limited extent. On balance, therefore, a relatively heavy burden is likely to fall on the lower income groups. Caution is necessary therefore to insure that those who call the loudest for an increase in pollution control are not just those who will be least affected or who can best afford a reduction in their standard of living if one solution is to call for a halt in the growth of GNP. Lower income groups generally are likely to take a dim view of a program that will not only cut their present consumption, but curb their hopes for future improvements. By no means is this meant to suggest that efforts to prevent environmental disruption should be halted. But it is necessary to remember that such a program will be expensive and that all will have to bear the costs. Unfortunately there seems to be no other alternative.

CONCLUSION

By this time it should be clear how complex are the problems of pollution. As our society continues to grow and expand, the control of pollution will become even more complicated. It is fortunate therefore that responsible citizens are becoming aware of the difficulties involved. Unless we have a running start, we may never catch up with the solutions that will be required. Some authorities argue that it may already be too late and that places like Lake Erie may be lost beyond hope.

Some biologists have suggested that we would have no problem today if our population had only stopped growing in 1850. They feel that if our population were only 23 million sprawled out across the whole continent, there would be so much water and air per capita that all of our wastes could be easily absorbed and diluted. At the same time there would be fewer wastes to contend with since not only would there be fewer people, but presumably there would be less need for some of the more hazardous products that have been necessitated by the increasing crush of population. Thus it is argued that there would be no need for insecticides, fertilizers, and pesticides since we would use a much more land-intensive form of farming and let nature do its own fertilizing.

Such thoughts are intriguing. To the economist, however, it does not appear that simple. Acknowledging that it is just as hard to prove, it nevertheless would seem that not only would the population have to stand still at the 1850 level, but so would technological and sociological change. If there were a smaller population in the United States, would it necessarily follow that the newer plastics and nonreturnable aluminum cans would not have been invented? Would we or the rest of the world have bypassed the discovery of atomic energy with its radioactive wastes? Would the country's population have remained predominantly rural or with time would there have been a desire to cluster in the cities? Assuming that people would eventually prefer to live in metropolitan areas, here would still be urban pollution (though perhaps on a smaller scale) and farmers would probably still want to use pesticides, herbicides and fertilizers.

Perhaps it is idle to conjecture about what could have been. Yet here are lessons to be learned from such speculation. As long as our technology and population continue to grow, the problems of pollution are likely to become more and more serious. It is more essential than ever that the population be alerted and that technology be harnessed for the prevention and treatment of pollution. This can and must be done if the affluent are not to choke on their effluent.

PART 1

THE NATURE OF THE PROBLEM

The great and dirty lakes

GLADWIN HILL

This article serves as an appropriate introduction to the dimensions of the problem of water pollution in the United States. The processes by which the Great Lakes have become polluted are typical of the recent history of many bodies of water in industrial areas throughout the country. The author is chief of the Los Angeles Bureau of the New York Times, *and has written extensively on the effects of air and water pollution.*

Whatever a honeymoon visit to Niagara Falls was like in the days when Blondin was crossing on his tightwire, it's different now. Something new and unpleasant has been added. Sightseers boarding the famous *Maid of the Mist* excursion boat are likely to find themselves shrouded in a miasma that smells like sewage. That's what it is—coming over the American falls in the Niagara River and gushing out of a great eight-foot culvert beneath the Honeymoon Bridge. As the little boat plows through the swirling currents to a landing on the Canadian side, it has to navigate an expanse of viscous brown foam—paper-mill waste out of the culvert—that collects in a huge eddy across the river.

While aesthetically shocking, this annoyance is hardly factually surprising. The falls in effect are the funnel through which the water from four of the Great Lakes passes on the final leg of its trip to Lake Ontario, the St. Lawrence River, and the sea . . . .

The Great Lakes constitute the largest reservoir in the world, containing about 20 percent of the fresh water on the face of the earth. They are the principal water source for one of the nation's largest concentrations of population and industry. If their waters become corrupted, it

Reprinted with permission from *The Saturday Review,* October 23, 1965, pp. 32–34.

would be a calamity of unprecedented magnitude. And it would involve not one but two nations, for the Canadian boundary bisects all the lakes except Lake Michigan.

If you stand on the shore at Duluth, Minnesota, and look out over the endless expanse of Lake Superior, it seems impossible that such a vast body of water could ever become tainted. Yet the flow of industrial pollutants alone from Lake Superior into Lake Huron has been measured by the International Joint Commission in hundreds of millions of gallons a year. The contamination gets rapidly worse as you move eastward along the chain of lakes to Michigan and Erie . . . .

Chicago, at the southern tip of Lake Michigan, has spent more than half a century and billions of dollars developing a good water system. The city draws a billion gallons a day from the lake, to serve 4,400,000 people. Its sewage, treated to remove most of the pollutants, is channeled southward into tributaries of the Mississippi, so that it does not affect the lake.

But around Chicago, extending past the Indiana line only ten miles to the southeast, is a network of small, sluggish waterways—the Grand Calumet River, the Little Calumet River, Wolf Lake, and various canals —that serve as a drainage system for a dense industrial complex sprawling for more than twenty miles along the shore, from Chicago through Hammond, Whiting and Gary, Indiana. The complex includes ten steel mills, five oil refineries, and dozens of other plants ranging from paper mills to soap factories. Six major plants discharge a billion gallons of waste a day that includes 35,000 pounds of ammonia nitrogen, 3,500 pounds of phenols, 3,000 pounds of cyanide, and fifty tons of oil. A good deal of this finds its way into Lake Michigan. There it has spoiled some of Chicago's best beaches, exterminated much aquatic life and recently defied city water officials' best efforts to provide a supply free of objectionable tastes and odors.

Reporting on a typical incident, Chicago alderman Leon Despres said, "In December 1964 we found millions of mysterious polyethylene pellets washed up on our two miles of breakwater. We learned that they represented just one flushing from a chemical plant, and that on the Michigan shore part of the same flushing made up thirty or forty miles of windrows."

"The southern tip of Lake Michigan," comments Wisconsin's Senator Gaylord Nelson, "is turning into a cesspool." He is concerned because similar troubles are developing one hundred miles up the west shore of the lake at Milwaukee.

Last March the Department of Health, Education and Welfare initiated, as it had in thirty-four previous major pollution situations across

ne country, a formal abatement proceeding for the Chicago area, with
llinois and Indiana as parties. In conference with representatives of the
·ublic Health Service, officials of the two states agreed to a corrective
·rogram based on detailed studies and recommendations of federal in-
·estigators. The program, which is enforceable by federal court action,
·rescribed waste treatment methods and standards to be instituted within
year.

Pollution from the western Great Lakes is augmented to a repulsive
·vel as their waters come down through the Detroit River—actually an
·terlake strait—and flow into Lake Erie.

"These waters," the Public Health Service summarizes, "are pol-
·ted bacteriologically, chemically, physically and biologically, and con-
·in excessive coliform (intestinal bacteria) densities as well as excessive
·uantities of phenols, iron, ammonia, suspended solids, settleable solids,
·hlorides, nitrogen compounds and phosphates."

Michigan has prided itself on enforcement of its pollution regula-
·ions throughout most of its 11,000 lakes and 36,000 miles of streams. But
·etroit, with its great industrial complex and a surprising horse-and-
·uggy municipal sewage system, has long been a stumper.

Apart from relatively minor contributions from Canadian commu-
·ities just across the strait, the Detroit area dumps twenty million pounds
·f contaminant materials into Lake Erie every day, in a waste flow total-
·ng 1.6 billion gallons.

About two-thirds of this is from industry and one-third is municipal
·ewage. Detroit gives the sewage of three million people only "primary"
·reatment, which means just the settling-out of grosser solids; standard
·ewage treatment today includes a "secondary" stage of chemical and
·iological neutralization of up to 90 percent of the contaminants.

Late in 1961 Michigan asked for federal help in cleaning up the
·etroit situation. The Public Health Service spent two years on an ex-
·austive investigation of pollution sources. The report named some of the
·ation's leading automobile, chemical, and paper companies as major
·ffenders. A cleanup program similar to the Chicago one was arrived at
·n hearings this spring.

Michigan waters account for only about 1 percent of Lake Erie's
·urface area. Conditions like Detroit's are repeated in various degrees at
·oledo, Cleveland, Erie, and Buffalo. No one knows the total amount of
·ollution pouring into the lake. But there is plenty of other evidence,
·oth visual and scientific, that the aggregate pollution has long since
·assed the danger point. It has transformed the 240-mile-long lake, Sen-
·tor Nelson remarked, "from a body of water into a chemical tank."

Because documenting interstate pollution, as the basis for involun-

tary federal intervention, is laborious, the Public Health Service had re
signed itself to proceeding from one metropolis to another in voluntar
intrastate actions, as with Detroit. But by this spring, public disgust a
the lake's deterioration had reached the overflow point. A one-man cru
sade by David Blaushild, a Cleveland automobile dealer, alone elicite
hundreds of thousands of signatures of Ohio citizens on protest petitions.

Feeling the heat, Ohio's Governor James Rhodes in June formall
requested the federal government to initiate an interstate abatement ac
tion. This automatically involved Michigan, Ohio, Indiana (which con
tributes pollution via the Maumee River, running through Ohio), Penn
sylvania, and New York. Two weeks of federal hearings in Cleveland an
Buffalo last August laid the groundwork for a cleanup program. It wil
cost upward of $100 million just to bring Detroit's sewage system up t
standard. The outlay confronting industry there for adequate waste treat
ment facilities is probably several times that. Along the whole lakeshore
comprehensive remedial measures will probably run into the billions.

These programs, under federal law, are not just pious declaration
of intent. The Secretary of Health, Education and Welfare is empowere
both to prescribe performance schedules and to reconvene the partie
customarily at six-month intervals, to review progress. At these session
the Public Health Service regularly gets impressive moral and technica
support from other federal agencies, such as the Fish and Wildlife Serv
ice, and from militant lay organizations such as the Izaak Walton Leagu
and the League of Women Voters, which has made water pollution re
form one of its major national projects.

There are those who gainsay the severity of Lake Erie's pollution
Mainly they are state and local water and health officials parrying im
plicit criticism of their activities and defending their areas against stigma
Life would be simpler for everybody if their position were substantia
But it is negative. They are arguing matters of degree. They can't an
don't contend that conditions are getting any *better* or that there's an
prospect, without action, of their getting anything but worse. Their argu
ment against corrective steps is that "we have no assurance of what they'
accomplish." This is rather like arguing against transferring from a sink
ing ship to a lifeboat on the ground that one doesn't know where the life
boat is going.

Pollution abatement officials concede that they cannot tell to wha
extent remedial measures can restore Lake Erie to its pristine cleanliness
The accumulation of contaminants down the years has disrupted its bio
logical metabolism to the point where it has the limnological equivalent o
cancer. Oxygen-absorbing chemicals have so sapped its supply of fre
oxygen, essential to the normal plant-fish-insect life cycle, that there is

,600-square-mile patch in the middle of the lake, equal to more than one quarter of its area, where the water for up to ten feet from the bottom is devoid of oxygen. The lake's general oxygen deficiency has set in motion a devolution of fish life from desirable species toward primeval anerobic sludgeworms and fingernail clams. Whitefish and pike, once the basis of a multimillion-dollar fishing industry, have vanished; in their place are inferior species such as carp that require less oxygen.

The disruption of normal processes has been aggravated by a runaway growth of algae, the plant life that in a healthy lake is microscopic; in Lake Erie it grows like seaweed, in great swatches up to 50 feet long. The dominant type is *cladaphora,* a coarse growth that looks as if it had dripped off the Ancient Mariner and smells terrible.

Algae growth is promoted by phosphates, which come particularly from municipal sewage. Detergents are composed up to 70 percent of phosphates. Millions of pounds of waste detergent pour into the lake every year. Each pound will propagate 700 pounds of algae. The algae absorbs more oxygen. When it dies, it sinks to the bottom as silt and releases the phosphate to grow another crop of algae. Copper sulfate will kill algae in a small volume of water, such as a pool. But biologists know of no chemical that can exterminate it on a scale such as is found in Lake Erie.

Even if the flow of pollution into Lake Erie were stopped entirely tomorrow, the reversal of this accelerating deteriorative cycle would be problematical. "All we can do," the scientists say, "is try."

Lake Erie contracted this disease rapidly because it is quite shallow. Its maximum depth of 210 feet is only one-eighth that of Superior, the deepest of the lakes. But Erie's fate obviously is simply a preview of what awaits all the lakes if pollution is not stemmed.

As Erie's turbid waters flow into the Niagara River, they pick up a final spate of pollution from around Buffalo. The Buffalo River is another industrial and municipal sewer, so polluted that it will not even support anerobic creatures: slime samples pulled up from the bottom are lifeless. The coliform intestinal-bacteria count recurrently exceeds 1,500,000 per 100 milliliters—1,500 times the safe level for human contact.

The conspicuous streaks that visitors a few miles away see in the water as it pours over Niagara Falls are described by Public Health investigators as "high phenol concentrations, oils, and high coliform counts [in a] polluted zone which extends about 400 feet from Bird's Island in the vicinity of Buffalo's sewage treatment plant, and widens to about one third of the river's width near Strawberry Island." . . .

Laymen aware of the Great Lakes' pollution, and their binational status, are often moved to suggest a massive international attack on the problem. This has logic but is impractical. Most of the pollution come from the American side. The International Joint Commission, which dea with boundary problems, is inherently cumbersome and slow. Its basi function is to advise the foreign affairs divisions of the two government and it can only take up matters formally referred to it after lengthy cor sideration on both sides. It took eight years of prodding by New Yor governors for the Lake Erie-Lake Ontario situation to be referred to th commission, and another two years before commission committees fo mally went to work on the matter last spring. Meanwhile, Canadian pro vincial authorities have been pursuing pollution abatement program roughly paralleling United States efforts.

"The crux of pollution abatement," observes Murray Stein, th Health, Education and Welfare Department's enforcement director, "i establishing individual pollution sources, defining corrective steps, an then following through to see that they're carried out. The Great Lake situation is the biggest problem we've ever tackled. But the importan thing is that it has now been tackled, and with reasonable cooperation i should be possible to clean it up. In the meantime, we've just got to hop for scientific advances that will make possible the undoing of the damag that's been done already."

Air pollution

ALLEN C. NADLER AND GLENN L. PAULSON

Allen C. Nadler, Assistant Clinical Professor of Medicine at Stanford University, is the author of the first sections of this article, which discuss The Atmosphere *and* Biological Effects of Air Pollution. *He offers a general survey of effects of air pollution on the body and an analysis of what a scientist means when he speaks of air pollution. Glenn L. Paulson is a graduate fellow in the Department of Environmental Biomedicine at the Rockefeller University. He concentrates on the effects of the Air Quality Act of 1967 and the role of the automobile in air pollution.*

THE ATMOSPHERE

Man's atmospheric environment is both narrow and finite; comprehension of its limitations and normal conditions is necessary to understand how it became polluted. The density of the atmosphere decreases with altitude, and approximately half of the atmosphere by weight lies below 18,000 feet. It contains about 21 percent oxygen which animals, including man, require for life and, because the average person requires available oxygen at pressures approximating 3 pounds per square inch, man cannot survive for long if oxygen is not available in close to that proportion and at that pressure. Other constituents of air include variable amounts of water vapor, nitrogen (78 percent), and carbon dioxide, carbon monoxide, and certain other gases, all of which total less than 1 percent by weight. The proportions of the gases are about the same in all parts of the world. The water vapor (water in a gaseous form) amounts to 1 to 3 percent by volume throughout the world's atmosphere. For our considerations, th water vapor can be regarded as an independent gas mixed with air.

There are several atmospheric layers. The *troposphere* is the laye adjacent to the earth and varies in height from about 28,000 feet over th poles to 55,000 feet over the equator, the depths being subject to sea sonal change. Normally, tropospheric temperature decreases with ir creasing altitude and we term this phenomenon the lapse rate. Where a abrupt change in the rate of temperature fall with altitude increase oc curs, we reach a region called the *tropopause*. This atmospheric regio separates the troposphere from the *stratosphere* (26 to 29 miles thick Our discussion will principally be concerned with the effects of man's ac tivity on the troposphere and, to some degree, the stratosphere.

The atmosphere is influenced by many forces, both natural an man-made. Chief among these is heat energy from the sun. Heat is a forr of energy as well as an expression of molecular activity. Since air is com posed of atoms and molecules, air temperature is also a measurement o heat or molecular activity. Because different materials have differen molecular structures, they will develop different temperatures (molecula activities) when the same amount of heat is applied. Accordingly, sub stances are said to have different specific heats or heat capacities. Land for example, becomes hotter than water when identical amounts of hea are applied and cools faster than water, as at night. Heat to the earth i largely supplied by the sun and this incoming radiation is offset roughl by outgoing or reflected radiation (terrestrial radiation). At night, coolin occurs by terrestrial radiation. Temperatures of land masses rise and fal more rapidly than water masses, and, therefore, the land is warmer b; day and cooler by night than the sea. This results in breezes toward lanc in coastal regions during the day which often reverse at night.

Atmospheric pressure is the force exerted by the weight of the at mosphere on a unit measurement of area (example, per square inch). W measure this force with an instrument called a barometer, one form o which is an evacuated tube with its open end placed vertically in an opei container of mercury. At sea level, the weight of the atmosphere acts as ; force on the mercury causing some of it to rise as a column in the tube on the average about 29.9 inches. Mercury is used rather than lighte substances such as water because the displacement of a heavier substanc in terms of column rise is considerably less, thus requiring a shorter tube Differences in atmospheric pressure between points on the globe account among other forces, for movement of air from regions of high to low pres sure. When this movement is parallel to the earth's surface, we refer t it as *wind*. Other factors responsible for air movement include the earth' rotation about its axis, its yearly revolution about the sun, the unevei heating of the earth's surface by the sun, and the tilt of the earth's axis

o mention a few. However, solar energy is the predominant force re-
ponsible for weather phenomena.

In discussing the tropopause earlier, we defined it as the altitude
one where an abrupt change in the lapse rate occurs. Actually, lesser
:hanges occur quite frequently and even closer to the earth. For exam-
ble, there may be a narrow layer within the troposphere in which tem-
)erature *increases* with altitude for several hundred feet; we call this an
nversion (or temperature inversion) of the usual decrease of temperature
vith altitude. Inversions can, thus, impede the rise of the air below and
f the latter air contains impurities (pollutants) the inversion acts as a lid
o seal them below. If no significant lateral movement of air (wind) oc-
·urs, then the stage is set for an acute air pollution episode in the volume
•f air below.

If the earth did not rotate on its axis one might conceptualize air
novement occurring by another means. We could conceive of warm air
ising over the equator where it is more heated and less dense. It would
ise high and flow laterally resulting in atmospheric pressure below lower
han that in the surrounding adjacent area where the air is more dense
)ecause of its cooler temperature. Thus, cool air from the poles would
nove toward the low equatorial pressure where it in turn would be
varmed, rise, and spread laterally toward the poles in a continuing
·ycle. For example, under these conditions, a lighter than air balloon
urned loose over the equators would rise and drift toward either pole. It
vould then descend over the poles and simply skim the earth's surface
oward the equator thence to rise again. However, the earth *does* rotate
ind this rotation results in a force which deflects the southern winds to-
vard the east in the northern hemisphere and toward the west in the
outhern hemisphere; we call this the Coriolis force. Other influences on
ocal weather include differential cooling and heating between mountains
ind flat land, desert and cultivated land, green and pavement, etc. All
·ontribute to weather phenomena.

Mixing Pollutants with the Atmosphere

We should regard the atmosphere as a gigantic reaction-vessel in
vhich countless largely unidentified chemical reactions take place, a reac-
ion-vessel whose properties change as temperatures and pressures
·hange, a truly dynamic state.

However, we have some knowledge of the basic constituents of the
itmosphere and of atmospheric chemistry. In addition to the oxygen, ni-
rogen, water vapor and oxides of carbon mentioned earlier, "pure" air
ncludes minute amounts of nitrogen and sulfur.

The oxygen-carbon dioxide cycle is of fundamental importance to animal-plant relationships. Atmospheric oxygen is produced by photo synthesis, a process by which plants exploit solar energy and trap carbon dioxide to synthesize organic (carbon-containing material) matter; a by product of this process is oxygen. Photosynthesis has been taking place for millions of years, and must continue in order to maintain the oxygen content of the atmosphere. While plants thus get their energy for growth and reproduction from the sun, animals extract energy from chemical packets (plants or other animals). In essence, animals burn material in the presence of oxygen is oxides of fuel materials, for example, carbon catalysts called enzymes and coenzymes. The result of this combustion in the presence of oxygen is oxides of fuel materials, for example, carbon dioxide. When man burns other materials in manipulation of his environment, other oxidized forms are generated, such as oxides of sulfur and nitrogen, etc.

Carbon Monoxide

The complete combustion (oxidation) of carbon in the presence of oxygen results in the development of carbon dioxide (CO_2). Carbon monoxide (CO) results from incomplete combustion and is an almost exclusively man-made pollutant. Its chemical fate in the atmosphere is not certain, although it may include conversion to CO_2, reaction with certain reactive chemicals (hydroxyl radicals) or assimilation by plant life on land or ocean absorption and biological oxidation. It is toxic to humans at concentrations of 100 parts per million with exposure for several hours. Carbon monoxide is produced almost entirely from incomplete combustion of fuel and predominantly by automobile engines, but tiny amounts are normally produced by man and other animals.

Carbon Dioxide

Carbon dioxide (CO_2), on the other hand, occurs naturally as a by product of animal respiration. However, a major contribution results from the combustion of fossil fuels such as coal, oil, and gas. Since the advent of the industrial revolution, man has increased CO_2 emissions, principally by electric power plant and internal combustion engines. It has been estimated (by Rohrman, *et al.*, in *Science* 156:931, 1967), that man-made emissions of CO_2 will show an eighteen-fold increase from 1890 to 2000. Because of pollution from these major sources, carbon dioxide enters the air faster than the natural cycle of carbon dioxide can adjust to the accelerated input. Besides assimilation of carbon dioxide by both aquatic

nd terrestrial plants, the oceans provide a huge sink for CO_2 with ex-
hanges taking place at air-water interfaces and shallow and deep water
ixing. Limestone (calcium carbonate) is formed by hydration of CO_2
nd reaction with calcium ion. However, the movement of CO_2 from
hallow to deeper parts of the ocean occurs slowly and thus is one of the
ate-limiting parts of the cycle of release into air, diffusion into water, etc.
his is one reason for the slow but measurable buildup of carbon dioxide
n the atmosphere.

Carbon dioxide has an effect on global temperature because of its
bility to absorb heat energy (infrared radiation) and reflect energy back
o earth much as the glass on a greenhouse does, allowing less heat to
scape to outer space. Since short wavelength energy (ultraviolet) passes
hrough carbon dioxide, an increase in CO_2 concentration with a more
r less constant supply of solar energy should result in increasing global
emperatures. Many scientists attributed the rise in mean global tempera-
ure in the 60 year period after 1880 to the seven percent increase in at-
ospheric carbon dioxide concentration. A decrease in mean global tem-
erature has been noted since 1940, and may be due to increases in water
apor and atmospheric turbidity from air pollution, resulting in less in-
oming energy reaching the earth, since they exert their own greenhouse
ffect in a reverse direction. Changes in solar activity could be another
ource of this decrease in temperature.

Other Weather-Changers

Indeed, various climatological changes are increasingly being at-
ributed to air pollution. Dust particles (800 million tons produced world-
ide per year) are effective cloud forming agents since their attraction
or water vapor permits condensation and ice crystal formation on the
ust nuclei. With sufficient moisture present, the droplets grow and fall,
esulting in the increased precipitation potential, rainfall, and weather
hange already believed to have occurred in many locales. For example,
ne investigator noted an apparent increase in cloudiness along the heav-
y traveled jet aircraft corridors between New York and Chicago; a
ound and one-quarter of water is released per gallon of jet fuel burned.
t altitudes of 20,000 to 40,000 feet the rarified atmosphere and cold
emperatures permit large volumes of air to be saturated by relatively
mall quantities of water. In addition, particulate exhaust from the jets
ontributes to the "seeding potential," setting up good conditions for
loud formation. Thus, even a seemingly innocuous substance such as
ater vapor can, under certain conditions, invoke profound environ-
ental change.

Nitrogen Oxides

Oxidation of nitrogen and release of nitrogen oxides into the at mosphere comes largely from automobile and electric power plan sources, as shown in the table on page 88. Man-made sources do not ac count for all or even the largest part of global nitrogen emission; othe sources include biological processes and lightning. Another source sug gested by Dr. Barry Commoner is the possible contribution of the ap plication of nitrate-containing fertilizer.

Some of the oxides of nitrogen are oxidized further to nitrogen diox ide which strongly absorbs ultraviolet light from the sun, creating nitri oxide and atomic oxygen (O); the latter can form ozone (O_3) in the pres ence of molecular oxygen (O_2). This highly reactive form of oxygen ha been responsible for ozone "alerts" in Los Angeles. During an ozon alert, out-of-doors exercise by school children is restricted in order t avoid lung damage. These ingredients in the presence of hydrocarbor (compounds containing carbon and hydrogen) react to form an eye-irr tating, mucous-membrane damaging substance called peroxyacyl nitrate or PAN, a principal component of photochemical smog. Other sources c hydrocarbons are natural bacterial decomposition of organic matter an release from forests, so that photochemical smog, while common in mos of our major cities, has the potential of forming even in rural areas give the presence of nitrogen oxides and sunlight.

Particulates

Particulate matter is enormously widespread. In New York City dust fall levels as high as 30 tons per square mile per month were re corded in 1969. Dust fall tends to consist of relatively heavy particle which settle close to their source. Finer particles settle at greater dis tances. Particles larger than 10 microns emanate from grinding, spraying and erosion. Particles less than 5 microns in size may reach the lowe respiratory passages and lodge in the tiny air sacs which terminate then Here there are none of the defenses which tend to expel the larger pai ticles from the upper respiratory passages. Sulfur dioxide may be at sorbed on those particles, and slowly released into the air sacs. Othe harmful components such as radioactive isotopes of polluted air can als be carried into the lungs on these tiny particles. Other aerosols, such a DDT, are blown from the forest, field or garden where they were orig nally sprayed and passed through food chains from plant to animal to ma or even directly to man. In this way, DDT has contaminated animal (in luding human) fat tissue in virtually all areas on the globe including the polar zones. Indeed, DDT has made its way into mothers' milk, resulting in the mother contributing to the contamination of her infant.

As indicated earlier, particles can scatter sunlight, reducing the amount of energy reaching earth and provoking global temperature decreases. Global dust levels are rising in the atmosphere and include products from all phases of human activity, including fossil fuel burning, and natural products such as volcanic eruptions and forest fires. Dustfall can also come from thermonuclear explosions which have radioactive isotopes as additional by-products.

Sulfur Oxides

Oxides of sulfur are emitted largely from man-made sources and are primarily released as sulfur dioxide, which may be oxidized to sulfur trioxide. The latter can combine with water vapor to form sulfuric acid mists which are highly corrosive to building materials, including stone and marble. Indeed, the oxides of sulfur can, in several decades, deteriorate statuary that has survived the ravages of several thousand years of normal weathering. When precipitated into water via rainfall sulfate salts increase acidity and can destroy aquatic life, most of which cannot survive at pH's (an index of acidity) lower than 4.0. Salmon kills occur at pH 5.5. While pH of lakes and rivers vary with seasonal changes in temperature and biological activity, the threat from sulfur oxides is formidable. Natural sulfur oxide sources include volcanoes, geysers and decomposition of sulfur-containing organic material.

Other emissions deserving mention include lead emanating largely from gasoline burning but also from smelting and certain glazing operations. About 65 percent of the lead in combusted gasoline is released into the atmosphere and results in both local and distant fallout. It is estimated that the northern hemisphere contains a thousand-fold surplus of lead above and beyond natural base levels because of man's contributions.

Radioactive pollutants vary from short-lived isotopes, such as radioiodine with an 8-day half life to radiocarbon with a 5,000-year half life and many intermediates. Those entering the environment as a result of nuclear explosions are discussed in *Environmental Effects of Weapons Technology* and *Nuclear Explosions in Peacetime;* those that come from nuclear reactors in *Environmental Cost of Electric Power*—all workbooks in this series.

Obviously, a variety of compounds are released into the atmosphere from all of man's activity with little regard for the how, when, or why of release. This anarchy of pollution results in the formation of many poorly understood or unidentified compounds, the short and long range biolog ical effects of which are even less well understood. For example, the com bustion of many plastic containers can result in the formation of totall new products reacting chemically in the atmosphere in ways completel foreign to us and producing effects which are as yet undetected or a least unknown. Polychlorinated biphenyls, used as plasticizers in th manufacture of many products, are chemically similar to DDT, and ar also appearing in the environment.

BIOLOGICAL EFFECTS OF AIR POLLUTION

In any metropolitan area, acute air pollution episodes can occur when ever atmospheric conditions prevent rapid dispersal or dilution of th pollutants. Acute air pollution episodes resulting collectively in the death of thousands in Belgium's Meuse Valley (1930), Donora, Pa. (1948), Lon don (1952, 1959, 1962), New York City (1953, 1962, 1966) have been wel documented. All of these episodes shared certain characteristics:

> A high population density with a correspondingly high concentra- tion of combustion processes.
> Seasonal influence—occurrence was in the winter when fuel con- sumption was increased and upper respiratory diseases were prev- alent.
> A stagnant air situation and temperature inversion for several days to a week with accumulation of pollutants in the air.

In general, the fatalities and severe illnesses resulted from acute chemically irritative changes to the lining of the air tubes (bronchi) lead ing to the lungs. In London, the leading causes of hospital admission during those episodes were respiratory disease, often with heart failur complicating the processes in the lung. The heart in a person with pre existing heart-lung disease could not take the added burden of movin blood through the chemically irritated lungs.

No single smog component in either London or Donora was presen in concentrations very much higher than usual. Thus, an increased dura tion of exposure is implicated, with the possibility of additive or syner gistic factors. (The latter is a biologic effect produced by two or mor agents together which is greater than the sum of the effects of the individ ual agents.) Other membranous body surfaces reflected the same irrita tive mechanisms, consequently, sore throat and burning of the eyes wer frequent complaints, as were headache and nausea. While no specific of fending agents have been indicted in the London and Donora disasters oxides of sulfur, common to both, were probably acting in concert with particulates and possibly other pollutants. Infecting agents may also have been operative. In Donora, 5,910 people were affected and 17 died. In London, 4,000 to 6,000 more deaths occurred between December 5–9, 1952 during a dense smog; in December, 1962, 340 more deaths occurred than was normal, during a similar period of smog.

For a long time the medical profession has been preoccupied with infectious causes of disease to the neglect of the physical and chemical aspects. While this was justifiable in the early part of the twentieth century when people were dying from diseases such as pneumonia, influenza, meningitis and tuberculosis, it no longer seems valid at this time in the United States when most people are dying from non-infectious diseases. Indeed, the diseases which appear to be killing Americans today seem not only to be non-infectious in origin but to have other common characteristics, namely multiple rather than single causes, insidious onset and development over a 20 to 30 year period and extremely difficult to treat when full-blown. Included in this group are cardiovascular disease, stroke, cancer, particularly bronchogenic cancer, and chronic pulmonary disease (bronchitis and emphysema).

Although complicated by many variables, the evidence linking chronic lung disease with air pollution is impressive. Bronchitis, an inflamation of the bronchi, is characterized by excessive mucous secretion accompanied by chronic or recurrent cough productive of sputum. Bronchitis is not considered chronic unless these manifestations are present on most days for at least three months of the year and for two successive years. Using these criteria, a 20 percent incidence is estimated in urban men in Great Britain between the ages of 40 and 60 years. In Great Britain, illness from chronic bronchitis is related to the population size of cities, suggesting that air pollution, which also increases with the size of cities, may be implicated. Another study showed a correlation between the bronchitis mortality rate and the amount of fuel burned for domestic and industrial purposes. In England and Wales, places with higher mean annual sulfur dioxide measurements had associated higher mortality rates due to chronic bronchitis. Using decreased visibility as an index to pollution, British investigators showed an association between pollution and illness-absenteeism among postmen. The postmen's counterparts working indoors suffered less work loss due to bronchitis, an association not attributed to weather alone. A number of observers have related aggravation of symptoms of bronchitis to air pollution increases.

Bronchial asthma, a disease in which the muscles of the bronchi constrict and impede outward movement of air, seems unquestionably to be influenced by air pollution. Thus, in Donora, 87 percent of asthmatics became ill, while illness struck 43 percent of the rest of the town's popu lation. Bronchial asthma generally responds well to certain medication which dilate the air tubes or combat allergy and infection. In contradis tinction, another type of lung disease, indistinguishable from asthma by physical examination but distinguishable by negative response to these medications has been described among American military personnel in Yokohama. Evacuation seemed to be the only form of therapy for "Yoko hama respiratory disease," pointing to some offending local environmenta contaminant. New Orleans has experienced epidemic outbreaks o asthma, typically in October. Dr. Murray Dworetsky, in the presidentia address to the American Academy of Allergy in 1969, pointed out tha "the literature strongly suggests that the frequency of death from asthm; has recently been increasing," and said that although "inappropriate man agement" is probably one cause, "There is much reason to believe tha in and of itself air pollution may be increasing the number of death from asthma."

Emphysema, another chronic respiratory disease, appears to be ad versely affected by air pollution. In this disease, the small air sacs into which the air passages empty become distended, rupture and/or coalesce The lining of these sacs is the site of gaseous exchange between air and blood. Thus, the larger sacs for the same volume present less surface area for exchange. Lung emptying is impeded, coughing is less effective, and victims become predisposed to infection. Heart failure is a common com plication. Emphysema prevalence in the United States is increasing; thi disease has doubled in incidence and mortality every five years for the past two decades. Not only is emphysema aggravated by air pollution especially in conjunction with smoking, but one California study showed that lung function could be improved simply by placing the patients in an air pollution-free room.

While most authorities agree that cigarette smoking plays the domi nant role in the development of lung cancer, there would appear to be some further agreement that an urban factor plays a much smaller al though detectable role as well. Data suggest a higher incidence of lung cancer in urban than in rural areas among smokers. Thus the incidence of lung cancer among smokers who emigrated from Great Britain to South Africa was higher than among white native South Africans who were even heavier smokers. Immigrants to Australia and New Zealand from Great Britain had higher lung cancer mortality than native New Zealand ers and Australians in spite of similar smoking backgrounds.

Air pollutants frequently have been concentrated and applied to the skin of mice. This has resulted in the

. . . induction of a kind of cancer called squamous cell carcinoma. Tumors under the skin have been induced in mice exposed to air pollutant tars collected from a number of American cities. The induction of lung cancer (adenocarcinoma) in mice has been achieved by exposure of mice in dust chambers to asphalt road sweepings and also by the same type of exposure to soot. . . . On the other hand rats and mice are quite resistant to induction of lung cancer. Even when exposed to aerosols of pure carcinogens, the animals do not develop lung tumors but do get skin tumors from this type of exposure although rats are normally resistant to skin tumor induction. Both species, however, develop lung cancers when the exposure is sufficiently intense. Thus, when the carcinogens are implanted in the lung in high dosages the animals develop lung cancer. . . .

It must be cautioned that such evidence is not definitive, although certainly suggestive. A one-pack-per-day smoker of cigarettes exposes himself to several hundred times more inhaled organic matter than an individual in a congested traffic area of New York. The major difference is that the smoker can stop smoking, while the pedestrian can hardly be expected to stop breathing.

New York City's 1953 episode was not really recognized until approximately nine years later when a comparative analysis of hospital records, and air pollution data revealed an excessive number of deaths during a period of severe air pollution. It is, therefore, hardly unreasonable to assume that many deaths from pollution may take place without ever coming to the attention of Public Health officials because of the unavailability of the necessary measurements or the necessary data analysis, or because the numbers are so small as to fail to give significant results.

Only three parts per million of sulfur dioxide exposure to a healthy person can produce a slight increase in airway resistance, that is, the ease with which expired air passes through the airways. Yet, people with chronic bronchitis or similar conditions may be aggravated by levels as low as 0.25. Twice as many acute respiratory illnesses were found at exposures to 0.25 ppm for twenty-four hours as at 0.4 ppm among those aged 55 or over with chronic bronchitis. Increased airway resistance can frequently occur even at extremely low levels of sulfur dioxide when it is combined with inhalation of particulate matter or sulfur trioxide, which combines with water to form sulfuric acid. Twelve-hour averages of sulfur dioxide (parts per million) in New York City have been known to exceed 0.8. The major source of sulfur dioxide in that city was the combustion of high sulfur content fuel oil and bituminous coal. The implication of these figures to New York's over one million estimated sufferers of asthma and hay fever could well be profound. Fortunately, some step have been undertaken to reduce the sulfur content of New York fuels.

Obviously, the respiratory system is most directly affected b breathing polluted air, and we have concentrated so far on cardio-resp ratory diseases in terms of possible cause and/or aggravation. Not i frequently, however, certain pollutants can attack an organ system fa removed from the portal of entry to the body. Lead, for example, ca have diffuse and confusing effects, as experience with occupational ex posure, and with exposure of children to lead paints has shown. Lead ca enter the body via lungs, intestinal tract, or skin and by poisoning certai enzymes which are present in most organs can affect blood-forming, nerv ous, gastrointestinal and excretory (kidney) systems among others. Lea toxicity can occur after chronic as well as after acute exposure.

The biological effects of carbon monoxide are different than man other air pollutants. First, it cannot be tasted, smelled or otherwis sensed by the body and second, it does not directly affect the eyes, nas passages or lungs. Instead, it passes unchanged through the walls of th lung into the blood, where much of it actively combines with hemoglobin the substance in the red blood cells normally responsible for carryin oxygen to all the tissues of the body. (A very small amount of CO i produced by normal body metabolism; we are concerned here only wit *additional* exposure to CO from the environment.) This combinatio forms a substance called carbomonoxyhemoglobin, and has the effect o decreasing the oxygen-carrying capacity of the blood. Since CO is abou 200 times more strongly bound than oxygen to hemoglobin, a smal amount of CO in the ambient air has a greatly magnified effect on th oxygen transport function of the blood. All tissues of the body may suffe from oxygen deprivation, but the two tissues most sensitive to lack o oxygen are the heart and the brain. Thus, at low levels, effects on thes two tissues are well documented. (At higher levels, about 1,000 ppm an more, CO can be lethal.) As Table 1 indicates, such effects can rang from changes in various psychological capabilities in humans, such a time discrimination, to permanent heart and brain damage in experimenta animals. (Concerning this last point, many scientists feel the now wel documented correlation between smoking and heart disease may well b due in part to the CO in cigarette smoke.)

The effects determined in the laboratory and described in Table should be compared with the actual measured levels of CO, culled fron several sources, listed in Table 2. The overlap is clear; the only uncer tainty concerns how long people in urban areas are exposed to thes various levels.

One of the air pollutants resulting from the aniline dye and benzen industries has, by careful epidemiologic analysis, been shown to result i

Table 1: Health Effects of Carbon Monoxide (CO)
(Prepared by D. M. Snodderly, Jr., New York Scientists'
Committee for Public Information)

Concentration of CO in air	% Carbomonoxy-hemoglobin in blood	Symptoms
Up to 300–400 ppm	30–40% and above	Severe headache, dim vision, nausea, collapse.[1]
100 ppm	Up to 20% depending on exposure and activity of subject	Headache at 20%. Impaired performance on simple psychological tests and arithmetic above 10% CO in blood.[1]
	20% in dogs exposed for only 5.75 hours per day, for 11 weeks	Brain and heart damage found at autopsy.[2]
50 ppm and below	2–4% and above Maximum of about 8% (calculated from [5])	Ability to detect a flashing light against dim background worsens with increasing amounts of CO. 4% was lowest point shown, but authors state that even the CO from a single cigarette could be shown to cause rise in visual threshold.[3] It is, therefore, obvious that smoking and exposure to CO from auto exhaust interact. Subjects presented with two tones and asked to judge which is longer. Judgment impaired at this level of CO in the air; lower levels of CO not studied.[4] Results interpreted as impairment of ability to judge time.[5] Not known whether this may influence people's ability to drive safely. Another author[1] concluded that 1–2% CO in the blood should cause a detectable number of errors on psychological tests if a sufficiently large-scale experiment were done.
15 ppm	Up to 2.4% (calculated from [5])	New York's air quality goal. Even this amount of CO could cause some of the effects on vision and loss of judgment of time that are mentioned above.

[1] J. H. Schulte, "Effects of Mild Carbon Monoxide Intoxication," *Archives of Environmental Health,* 7, 1963, pages 524–30.

[2] F. H. Lewey and D. L. Drabkin, "Experimental Chronic Carbon Monoxide Poisoning of Dogs," *American Journal of Medical Science,* 208, 1944, pages 502–11.

[3] R. A. McFarland, F. J. W. Roughton, M. H. Halperin and J. I. Niven, "Effects of Carbon Monoxide and Altitude on Visual Threshold," *Journal of Aviation Medicine,* 15, 1944, pages 381–94.

[4] R. R. Beard and G. Wertheim, "Behavioral Impairment Associated with Small Doses of Carbon Monoxide," *American Journal of Public Health,* 57, 1967, pages 2012–22.

[5] J. R. Goldsmith and S. A. Landau, "Carbon Monoxide and Human Health," *Science,* 162, 1969, pages 1352–59.

Table 2: *Carbon Monoxide Levels at Various Locations*

Location	CO Levels (in average ppm's)
Los Angeles Freeways	37
Los Angeles Freeways, slow, heavy traffic	54
Los Angeles, severe inversion	30 for over 8 hours
Parking garage	59
Cincinnati intersection	20
Detroit, short peak	100
Detroit, residential area	2
Detroit, shopping area	10
Manhattan intersection	15 all day long
Allowed industrial exposure for 8 hours (for comparison)	50 recently lowered from 100

the increased occurrence of bladder cancers among workers in these in dustries. In this instance, the target organ, the bladder, was certainly dis tant from the portal of entry, the lungs. Since bladder cancers are respon sible for only a few percent of all deaths from malignancy, these pockets of disease could be expected to attract attention, and did. The disease was found to be heavily concentrated in workers within these industries *and in residents in the immediately surrounding vicinity.* Laboratory con firmation by the provocation of experimentally induced aniline dye tu mors in dogs using beta-naphthylamine has been reported. An increased frequency of tumors of the bladder paralleling the increased occurrence of lung cancers in smokers has been reported and as one might expect cigarette smoke contains these same contaminants. In the case of this pollutant, short term exposure revealed no ill effects; it was exposure for a long time to a relatively low level that resulted in disease.

Similarly, the results of airborne radioactive isotopes are slow in developing. In 1954, fifty-three Marshall Islanders were subjected to radiation fallout including iodine 131 from a nuclear bomb test. Iodine 131, although a short-lived isotope, concentrates in the thyroid and can damage the thyroid tissue as it emits its energy and decays. Eleven and twelve years later, eighteen of these individuals were reported to have thyroid abnormalities. In eleven cases, surgery was performed, and one cancer of the thyroid was found. Fallout was not blamed for the malig nancy found in another patient, who was reported to have received less exposure from iodine 131. However, in a population of this size, one would not expect even a single case of thyroid cancer to be present in

ifteen years. It should be remembered that this represents also a rather
brief period between exposure and disease. The mutagenic properties of radiation are well known, and would lead us to expect effects on later genera-
ions. Experience has taught us that it may be as many as five generations
after exposure before the effects of a recessive mutation appear. These
concerns are probably quite applicable to the long term effects of release
of radioactive materials into the environment from whatever source.

There is much that we don't know about what pollutants are in the
air. As Glenn Paulson shows, the term "particulate matter" is a general
one that includes numerous pollutants, many as yet unidentified. There
are also gaseous pollutants emitted by various industries or released in
the burning of wastes that are not monitored. If we knew more about
what is in the air, this would be only the first step toward studying the
biological effects of single pollutants and pollutants in various combina-
ions.

Early concern with the effects of air pollution was largely confined
to effects on man which, while quite proper, was also somewhat misleading. By the time we started noticing damage in man, a devastating toll
had been taken in plants and possibly in lower animals. Long before the
health effects of air pollution became a matter of serious concern, enormous areas of formerly fertile ground surrounding ore processing mills in
this country and others became bare as a result of fumes emanating from
the mills. This occurred during the days prior to the development of high
stacks which promoted more rapid dilution. Sulfur dioxide from stack
gases tends to burn vegetation, particularly alfalfa and soft-leafed vegeta-
bles. Hydrofluoric acid has been described as being damaging to plants
and fluorosis has been described in cattle. By the end of the late 1940s
more and more complaints from farmers were heard concerning smog
injury to crops. Virtually every crop in New Jersey has been adversely
affected by air pollution.

The effect of smog on plants can be quite variable and consequences can include the reduction of crop yield, growth retardation or
outright destruction of the plant. Visible damage is probably best known
since it is the most easily observable, as when pine trees lose their nee-
dles, or when lawns become brown after an intense smog assault. Of continuing concern should be the fact that the more extensive the damage
to plants, the greater the reduction in oxygen production by green things.
For example, phytoplankton, microscopic marine plants which produce
oxygen, can be destroyed by air or water pollution by DDT.

City growth, increased miles of highways, and the spreading out of
more and more people over the countryside reduce the area given to
plants producing life-giving oxygen. Statistics on how fast land is being

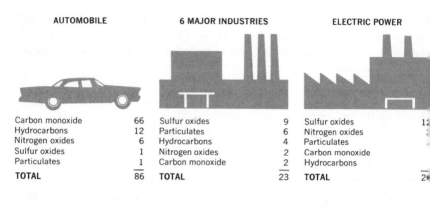

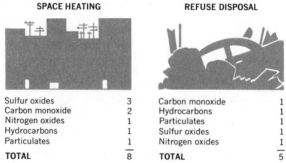

ANNUAL EMISSIONS OF FIVE MAJOR POLLUTANTS IN MILLIONS OF TONS, AS OF 1966
TOTAL: 142 MILLION TONS

POLLUTION AND ITS SOURCES
Five major pollutants and their sources are shown above. The total for
industry is obviously incomplete, since it includes only the six major
industrial polluters. Putting all pollutants into the same units—millions
of tons—is somewhat misleading. Some pollutants are harmful even
in very small amounts. Some are more harmful than others, and some
are more harmful together than separately. (From "The Sources of Air
Pollution and Their Control," Public Health Service Publication No.
1548, Washington, D.C., 1966.)

consumed are not available for many sections of the country. However,
according to the U.S. Department of Agriculture, roughly 420,000 acres
a year are being converted to urban use (which includes buildings and
roads), and approximately the same amount is going underwater as new
dams and reservoirs are built. In addition, approximately 160,000 acres
are being converted each year in rural areas for highways and airports.
It is estimated by the USDA that cropland furnishes roughly one-half of

his million acres. Thus man, with his exponentially increasing population and his soaring per capita energy use, at least in this country, may ultimately threaten one of his own basic life-sustaining systems, the oxygen cycle.

The measures taken so far to reduce air pollution are being offset by increases in the sources of pollution—more and larger power plants, more and larger industrial plants, more automobiles and trucks. So far, controls are being imposed on industry slowly, and principally to reduce sulfur dioxide and particulate matter. Knowledge of what comes out of industry's stacks is often limited, and authority to control it is even more limited. For example, the Committee for Environmental Information found that while industries reported to the city's Air Pollution Control officer what was coming out of their stacks, this was sometimes privileged information, not available to the public. The Air Pollution Commissioner could order reductions in industrial emissions only for sulfur dioxide, hydrogen sulfide, oxidants and particulates—the pollutants for which standards have been adopted by the city. He has authority to reduce emissions of other pollutants only if he can show that they present a danger to the health of the people in the vicinity or constitute a nuisance. Direct cause and effect on human health is extremely difficult to prove in connection with any environmental contaminant, and for new and untested chemicals would be impossible to prove until after the tragic fact. (The dye industry example described previously is a good case in point.)

At a recent conference on air pollution and the automobile at the University of Missouri William H. Megonnell, Assistant Commissioner for Standards and Compliance, National Air Pollution Control Administration, said:

> In my judgment, the best we can expect from the Federal standards now in effect is that hydrocarbon and carbon monoxide emissions will in 1980 dip to approximately 60 percent of current emissions, or roughly what they were in 1953. And after 1980, when these standards have passed the saturation point of their effectiveness, as vehicle use continues to increase, the levels of pollution will resume their upward climb.

At the same conference, Dr. Robert Karsh, president of the St. Louis Committee for Environmental Information, said:

> The 1968 automotive emission standards reduced carbon monoxide emissions by 50 percent and hydrocarbons by 70 percent of uncontrolled levels in new cars only. The devices are not maintained because they do not have to be maintained in most areas. Because of

increased numbers of cars and increased driving, under existing controls automotive pollution will double in the next 30 years.

Since that conference, new federal standards have been proposed which would reduce carbon monoxide emissions to half of present emissions, hydrocarbons to a fourth, particulates to a third and nitrogen oxide to a sixth. These stringent standards would not go into effect until the 1975 model year, and therefore would not reduce the total automotive pollution by those amounts, and then only if the control devices are maintained, and the number of cars does not increase.

According to the U.S. Bureau of Public Roads, the number of registered motor vehicles is increasing every year—in 1969 the increase was three million over 1968. But this tells only a small part of the story. The amount of air pollution from cars is more closely related to miles travelled, and particularly to *urban* miles travelled. In 1946, urban miles travelled were 170 billion. Twenty years later this had more than doubled, to 470 billion, and it is still rising. What this indicates is that our control efforts continue to lag behind our capacity to pollute the air.

A PIECE OF THE ACTION*

Other sections of this workbook document the known and potential effects of air pollution on human health as well as on the entire biosphere. This information, once understood by interested citizens, can cause more than just generalized concern, and may be applied in specific cases in several ways. In this section, two such applications will be discussed: the Federal Air Quality Act, at this writing useful for dealing with problems of sulfur oxides and particulate matter; and air pollution and urban expressways.

The Air Quality Act of 1967

The Federal Air Quality Act of 1967 (Public Law 90–148) is long and complex; the sections of greatest immediate use to citizens are titled "Air Quality Control Regions, Criteria and Control Techniques" (Section 107) and "Air Quality Standards and Abatement of Air Pollution" (Section 108). These portions of the Act do not *themselves* establish allowable levels for pollutants; instead they establish a mechanism which includes a complicated series of hearings and administrative steps. This procedure

*This section was written by Glenn L. Paulson.

llows any interested citizen, be he an elected or appointed government
official, an industrialist, a medical or scientific expert, or, most impor-
antly, *any* interested individual, the chance to state how dirty (or clean)
he wants the air he breathes to be.

The process works as follows: First, HEW designates "air quality
control regions based on jurisdictional boundaries, urban-industrial concentrations, and other factors . . . ." Many such areas are already designated. A typical example with its complete, if cumbersome, official title
is the "Metropolitan Philadelphia Interstate Air Quality Control Region,"
which includes: Bucks, Chester, Delaware, Montgomery and Philadelphia
Counties in Pennsylvania; Camden, Burlington, Gloucester and Mercer
Counties in New Jersey; and Newcastle County in Delaware. Other fully
designated areas are:

Washington, D.C.	Cleveland
Metropolitan New York	Pittsburgh
Chicago	Buffalo
Denver	Kansas City
Los Angeles	Baltimore
St. Louis	Hartford–Springfield
Seattle–Tacoma	Steubenville
Boston	Indianapolis
Cincinnati	Minneapolis–St. Paul
Louisville	Milwaukee
San Francisco	Dayton
Detroit	Providence

Among those in the process of designation are:

Chattanooga	Birmingham
Toledo	Houston–Galveston
San Antonio	Phoenix–Tucson
Dallas–Fort Worth	Memphis

There are still more scheduled for the future.

According to HEW, by the end of the summer of 1970 some fifty-
seven regions, including about 70 percent of the nation's urban residents,
will be formally designated. During this process, HEW develops "con-
sultation documents" which serve as the basis for setting the geographical
boundaries of the region. These documents, available to the public, serve
as one source of technical information on the condition of that region's
air.

A second aspect of the mechanism is the publication by HEW of
two separate technical documents on each specific pollutant that may be

affecting "the health and welfare" of the population. At the time of thi writing (February, 1970) only two such pollutants, sulfur oxides and par ticulate matter, have been so considered. In the future, documents wil be published on other pollutants, including carbon monoxide, the oxide of nitrogen, photochemical smog, lead, etc. The first of each pair of docu ments is called a "criteria document," *e.g.* "Air Quality Criteria for Sul fur Oxides." This is really an incorrect name. In fact the criteria docu ment is a summary of *facts*, not criteria or standards, from the scientifi examination of the effects of a particular pollutant (or a combination o pollutants) on man, other animals, plants, objects, and the atmosphere i general. For example, the sulfur oxides document includes chapters on Physical and Chemical Properties and the Atmospheric Reactions of th Oxides of Sulfur, Sources and Methods of Measurement of Sulfur Oxide in the Atmosphere, Atmospheric Concentrations of Sulfur Oxides, Effect of Sulfur Oxides in the Atmosphere on Materials, Effects of Sulfur Oxide in the Atmosphere on Vegetation, Toxicological Effects of Sulfur Oxide on Animals, Toxicological Effects of Sulfur Oxides on Man, Combined Effects of Experimental Exposures to Sulfur Oxides and Particulate Mat ter on Man and Animals, and Epidemiological Appraisal of Sulfur Oxides

The first published version of this criteria document was over 37(pages long. Fortunately, interested individuals need not wade throug all of this technical material. Summaries of the two published criteri documents are available from HEW, including a further distillation, o resume. The resume from the sulfur oxides summary is as follows:

> In addition to health considerations, the economic and aesthetic benefits to be obtained from low ambient concentrations of sulfur oxides as related to visibility, soiling, corrosion, and other effects should be considered by organizations responsible for promulga- ting ambient air quality standards. Under the conditions prevailing in areas where the studies were conducted, adverse health effects were noted when 24-hour average levels of sulfur dioxide exceeded 300 µg/m³ (0.11 ppm) for 3 to 4 days. Adverse health effects were also noted when the *annual* (our emphasis) mean level of sulfur dioxide exceeded 115 µg/m³ (0.04 ppm). Visibility reduction to about 5 miles was observed at 285 µg/m³ (0.10 ppm); adverse ef- fects on materials were observed at an annual mean of 345 µg/m³ (0.12 ppm); and adverse effects on vegetation were observed at an annual mean of 85 µg/m³ (0.03 ppm). It is reasonable and pru- dent to conclude that, when promulgating ambient air quality standards, consideration should be given to requirements for mar- gins of safety which would take into account long-term effects on health, vegetation and materials occurring below the above levels.

Each criteria document, as it is published, can be considered the best available compilation of the technical data *at the time of publica-*

ion. Future studies may add new data indicating, for instance, adverse effects on human health at even lower levels of pollutants. In some cases his has already occurred. At a conference held after the publication of he criteria document on particulate matter, a paper was presented in-licating adverse effects on the lungs of children at levels of particulate matter even lower than those stated in the document. (Presumably the criteria documents will be revised and new information incorporated at some time in the future.)

Space does not permit a full explanation of the contents of the criteria documents; interested individuals can obtain them from HEW. However, two points should be made explicit.

The first point concerns the actual concentrations of a pollutant said to be harmful, *e.g.,* "Under the conditions prevailing in areas where the studies were conducted, adverse health effects were noted when the annual mean level of particulate matter exceeded 80 micrograms per cubic meter." (Units are explained below.) There is still controversy over certain aspects of such statements. No one doubts that high concentrations of air pollution for several days (caused in many cases by specific weather conditions described earlier in this workbook) can cause increased illness and death in humans. Such events have occurred several times in London and New York City, as well as in Donora, Pa., the Meuse Valley, and other places. The more controversial question is whether long-term exposures to the lower levels of air pollutants found every day in urban areas have effects on health. As recently as 1965, the President's Science Advisory Committee said:

> While we all fear, and many believe, that long-continued exposure to low levels of pollution is having unfavorable effects on human health, it is heartening to know that careful study has so far failed to produce evidence that this is so, and that such effects, if present, must be markedly less noticeable than those associated with cigarette smoking. Attempts to identify possible effects of ordinary urban air pollution on longevity or on the incidence of serious disease have been inconclusive.

The studies cited in the criteria documents indicating that such effects *do in fact* occur contradict this statement. In part this is because much of the work described in the criteria documents has been published since 1965. Another aspect is the uncertainty inherent in such studies, many of which are epidemiological in nature (that is, the studies correlate changes in the numbers of people suffering particular health effects with changes in the amount of a particular pollutant in the air of their community). Moreover, that this controversy exists should, on reflection, not be surprising. The environment, after all, is very complex and not very well understood. The extent of our ignorance is in fact quite considerable in many areas. For example, if one examines federal data on levels of particulate matter in urban air, one finds the following:

Suspended particulates—102 micrograms per cubic meter ($\mu g/m^3$) average

Analyzed fractions	$\mu g/m^3$
Benzene soluble organics	6.9
Benzo(a)pyrene	.002
Ammonium	0.6
Nitrate salts	2.9
Sulfate salts	10.7
16 metals—at highest measured levels in U.S.	18.33
Total known	39.432 $\mu g/m^3$

Thus we know the exact chemical composition of less than 40 percent of the dirt in the air of our cities. Because of the way the data are given above, this is a somewhat generous estimate of our knowledge; for most cities, probably only 25 percent of the dirt can be specifically identified.

Given this largely unknown composition, and the resulting toxicological uncertainties, it is to be expected that epidemiological studies in different cities may associate adverse health effects with different average levels of particulate matter. Other factors—for example, size of the particles, climate, presence of different bacterial and viral organisms, interactions of the various pollutants, etc.—no doubt also enter into this lack of detailed compatability between studies. Nevertheless the most well-conducted studies all indicate strongly that chronic exposure to the low levels of air pollutants found in many urban areas is taking its toll on health. It should be stressed that in most cities, such medical studies have not yet been carried out.

The second point is that, even given the uncertainty in the absolute numbers used, and the possibility that in the future adverse effects on health will be demonstrated at even lower levels, the numbers in the criteria documents *are the technical basis* for HEW's review power over standards for these pollutants set by each state for the designated urban areas in its borders. Thus they are an integral part of the existing law, and must be recognized as such.

So far nothing has been said about the second of each pair of the HEW documents for every specific pollutant. These are similar in length to the criteria documents, but devote themselves to techniques for cutting down or eliminating the actual emissions of the pollutant. At this writing, the two already issued are titled "Control Techniques for Sulfur Oxide Air Pollutants" and "Control Techniques for Particulate Air Pollution." By this mechanism, HEW insures that a good summary of the techniques, both existing and potential, for controlling a pollutant is available at the same time standards are set for the amount of that pollutant to be allowed in the air.

(These documents, though essential to the implementation of the Air Quality Act, require some familiarity with engineering principles and will not be discussed further.)

When a state governor is confronted by these three items, namely, a criteria document, a control technique document, and an area in his state that is part of a designated air quality control region, he must move to carry out his responsibility under the Act. (This responsibility is usually allocated to an air pollution agency at the state level, though some states have different mechanisms.) It is at this point that all interested citizens, with technical data in hand and other considerations in mind, can speak their piece to influence the quality of the air they wish to breathe. The state must propose "air quality standards," that is, target goals describing the amount of the specific pollutant that will be allowed in the air, and then hold public hearings to determine whether these standards are acceptable to the people living in the area. These standards are phrased in somewhat technical language. For example, some of the standards originally proposed by the State of Pennsylvania were: Suspended particulates: an annual average of 100 micrograms per cubic meter with 80 as a long range goal. Sulfur dioxide: an annual average of .03 parts per million.

It is quite easy to understand these standards. Those for gaseous pollutants, such as sulfur dioxide, are often given as a yearly average (arithmetic or geometric mean) of so many "parts per million," or ppm. One part per million is exactly what it says: one part of something in a million parts of something else. For example, 1 ppm is approximately equivalent to 1 ounce of whiskey in a 9' x 12' living room, an amount which, if effectively delivered, can have an effect on a human being. Similarly, air pollutants can have effects at these exceedingly low levels. Suspended particulate matter standards, and sometimes sulfur dioxide standards, are formulated in terms of "weight per unit volume," usually "micrograms per cubic meter" or $\mu g/m^3$. (A microgram is one-millionth of a gram, which is in turn the weight of a cube of water about 0.4 inches on a side; a cubic meter is somewhat bigger than a cubic yard.) Often qualifications on the standards will be stated—for instance, the number of days per year or hours per year that a certain peak level can be exceeded. In most cases, these standards can be easily compared to the values in the criteria documents. (If comparisons are not easily made directly, conversions can be calculated by state or federal experts, or by other interested technically trained people.) For example, the original Pennsylvania proposal was for 100 $\mu g/m^3$ particulate matter (yearly average); the HEW criteria document on particulates stated that "adverse health effects" have been observed above 80 $\mu g/m^3$.

This comparison was not without effect on the citizens of Pennsylvania. According to an article in the *Wall Street Journal* (October 20, 1969) on the "breather's lobby," about 700 persons attended, with many testifying at the required public hearings in Philadelphia and Pittsburgh; the eventual standard for suspended particulates recommended by the state agency was 65 $\mu g/m^3$. This process of citizen testimony, based on an understanding of the technical questions involved, has been repeated in many cities across the country, and will undoubtedly occur in more, not only for sulfur oxides and particulate matter, but for other pollutants. It requires some homework and attention to technical matters. Many thousands of people interested in the quality of their air have already made this effort to use the scientific information available to them to begin to halt the decline in the quality of the air they breathe, and to start the process of improvement. This citizen awareness and involvement is a hopeful sign.

However, environmental improvement is not brought about by standard setting alone. As the *Wall Street Journal* article concluded, "In a sense, yesterday's events are only a preliminary triumph for the breather's lobby. Now the proposed standards must be forwarded to HEW, which must pass on them. Then comes the massive task of devising plans to implement the standards and setting a timetable for action."

The plans to implement the standards, that is, to reach the goals, and to set the timetable will require hard economic and social decisions, with many political, philosophical and moral judgments. For example, if expensive controls are required of electric power plants, how should this affect what the consumer pays? If the cost of electricity to the average household is increased by $3.00 a year in order to reduce utility emissions of sulfur dioxide by 90 percent (as Robert Kohn has estimated for the St. Louis area), is the improvement in air quality worth that cost? This is not a scientific question, but one that is essentially political and economic. More basic and far-reaching decisions may be required: Shall power be generated by burning coal, oil or gas, by using nuclear or hydroelectric power, or by some other means? Where shall power plants be located?

How fast shall we increase the amount of power generated and, indeed, should increases in power generation be permitted at all?

These considerations obviously go far beyond the question of a standard for one or another pollutant and, indeed, for many reasons it is necessary to ask whether or not the setting and enforcement of standards is, *in itself,* an adequate technique for dealing with the many-headed Hydra of environmental decay. It is usually implicitly assumed, and often explicitly stated, that standards are set at "safe" levels where the people exposed (or the environment insulted) will show no adverse effects. History has shown that this is too often not the case. Standards are usually a compromise, reached after an interplay of many factors. In fact, as time has gone on and more observations have been made, many "standards" for human exposure have been shown to be unsafe for certain segments of the population, as demonstrated by the continuous lowering of certain allowable exposure levels for industrial workers, for example. In fact, recent data from Japan mentioned above indicates that a particulate level of 80 μg/m^3 may not be low enough to completely protect the health of school children.

Technically, this is usually discussed in terms of whether or not there is a "threshold" below which no adverse effects occur. For certain environmental contaminants, radiation and carbon monoxide in particular, the best scientific judgment is that there may well be no threshold, or no "safe" level for human exposure above normal background levels. Thus any standard for these is a compromise based on the weighing of the benefits provided by the technology producing the contaminant against the risk to health from the pollutant. For some contaminants (such as airborne lead), debate rages heavily, providing much needed life to scientific meetings. But it will be apparent to all that a value judgment *not* confined to scientific data underlies these two different views: standards are considered by some as "licenses to pollute" and by others as "law to protect." This issue has no technical solution in general terms, but should be kept in mind by any citizen working to protect and improve the environment in which he lives.

The present approach to standards has other limitations. One is that taking pollutants one by one and at the same time allowing new and unknown pollutants to be added to the air has obvious limitations as a strategy for improving the overall quality of the air. Essentially what the present system does is to assume that any substance added to the air is acceptable until proven to be deleterious to health—"innocent until proven guilty"—a principle more appropriate to individuals at the bar of justice than to a pollutant at large in the atmosphere. Dr. Raymond

Slavin, presenting testimony on behalf of the St. Louis Committee for Environmental Information to the Subcommittee on Air and Water Pollu tion, Senate Committee on Public Works, said:

> It seems to us that this whole approach to air pollution and health is upside down. Why must we leave a pollutant in the air until it can be proven that it is harmful? Why should not the burden of proof be the other way around? That is, no one should release any substance into the air other people must breathe unless it can be shown to be *harmless,* unless he can prove that it is *not* injurious. This is the guiding principle in the acceptance of food additives and drugs. Proof must be submitted that a new food additive or a new drug will not harm human health before it can be placed on the market. While this system is a long way from perfection, the principle is sound, and perhaps even more appropriate to air. We do have some choice in the matter of the foods we eat and the drugs we use, while we have very little choice as to the air we breathe.

Not only do we presently permit the addition of new pollutants to the air, but we do little to limit the proliferation of sources of the old ones, allowing more and more automobiles and expressways, larger and larger cities, unrestricted growth of industry and power plants.

Air Pollution and Urban Expressways

Of all the pollutants in urban air, both known and unknown, the most common by far is carbon monoxide (CO). For example, a report from the National Academy of Sciences a few years ago states that of the approximately 125 million tons of pollutants produced that year in the United States, 65 million tons, or 52 percent, was CO. CO is formed when carbon in any fuel is incompletely burned so that instead of becoming carbon dioxide (CO_2), it stops half-way at CO. Thus CO can potentially be formed from any carbon-containing material—oil and its products, coal, natural gas (as in a home stove), wood and even tobacco. However, by far the most common source in the urban air is incomplete combustion in the internal combustion engine, primarily from cars. Of the 2,726,000 tons of CO produced in Chicago in 1966, for instance, 2,462,000 tons, or over 90 percent, came from cars, trucks and buses, with probably well over 95 percent of the total from cars alone.

Much more is known about the biological effects of CO than about any other pollutant in automobile exhaust. These other pollutants, all present in significant amounts, include unburned gasoline (hydrocarbons), the oxides of nitrogen, lead, carcinogens (cancer-causing chemicals including "polynuclear aromatic hydrocarbons" such as benzo(a)pyrene) and others. For an idea of the amounts, the concentrations of both the oxides of nitrogen and CO in cigarette smoke are very similar to those in automobile exhaust. Certain of these substances can undergo further reactions with each other in the atmosphere, especially in sunlight, to give rise to a whole series of new substances, ozone and peroxyacyl nitrates (PAN) for instance.

The connection between CO and urban expressways, or any main traffic artery, is apparent from the discussion of its biological effects on pages 84–86. The more cars, in general, the higher will be the CO concentration. There is one correction to this: in general, the faster a car goes, the *lower* the concentration of CO in the exhaust, but the *higher* may be the concentration of other exhaust components, such as the oxides of nitrogen and lead. But since a main purpose of an urban expressway is to increase the traffic volume (in vehicles per hour) as well as the average vehicle speed, a new expressway may possibly *increase* the total CO level in the air near the expressway. This prediction was in fact made in a preliminary engineering study by the New York City Department of Air Resources concerning the impact of the proposed Lower Manhattan Expressway on the air quality in that area. This study, admittedly crude because it was one of the first if not the very first of its kind, predicted CO levels, depending on road design and traffic volume, of from 50 up to 300 ppm or more. (Three hundred ppm breathed for a few hours may cause a person to fall unconscious.) The study did not deal with other pollutants.

The Lower Manhattan Expressway, which had first been planned as a surface and then as an elevated road, was being proposed in 1968 as a trench, depressed below street level, with buildings above all or part of it. A local planning board asked the city's Department of Air Resources whether it could create serious air pollution problems. The Department of Air Resources made the study mentioned above, and communicated it to several city agencies, but not to the public.

Word of the study got out, and pressure from citizens, the press and from Congressman Edward I. Koch, who represents the 17th District in Lower Manhattan, brought it into the open. Predictions in terms of ppm of CO, however, were meaningless to the public, and the New York Scientists' Committee for Public Information was asked for an interpretation. Their brief analysis, of which the data in the table on page 85 is a part, was given wide publicity.

The analysis and the subsequent public discussion had two important effects. First, opposition to the Expressway in the community, already strong, increased dramatically and, because of the air pollution aspect, involved individuals and organizations throughout the metropol-

itan area. (One new slogan, not strictly justified by the technical discus
sion, was "Kill the Expressway before it kills you.") The idea for a Lowe
Manhattan Expressway is now dormant, if not dead, and the air pollutio
potential of new urban expressways is currently being debated in many
cities around the country.

The second consequence, important to all urban areas, is a very de
tailed and extensive series of measurements, for the first time, of levels o
CO and other automotive pollutants including noise near roadways o
various shapes and traffic levels in New York City. This study, being
carried out by the Department of Air Resources with federal funding
will provide the first hard data on this question. Interest in the results o
this study is very great in view of the current notion of using urban ex
pressway locations not only for vehicles, but also for buildings above the
roads. This concept of double use, called "air rights utilization," is in
great favor among architects, city planners, and others. Yet, unless the
pollution potential of such construction is taken into account and mini
mized, the environmental impact on people working, living or studying in
such buildings may be tremendous. For example, a school, among othe
structures, was to be built over the Lower Manhattan Expressway. On
wonders what the impact on the children's learning capacities would have
been if CO had seeped into the building. That this is not idle speculation
is demonstrated by recently obtained federal data showing the CO level
in the neighborhood of the many-laned approach to the George Wash
ington Bridge and the apartment buildings astride it in New York. The
CO levels *inside* a third floor apartment averaged 14 ppm all through a
24-hour day. This can be compared with levels of about 15 ppm meas
ured at one corner in Manhattan during the busiest part of the day
(9 a.m. to 7 p.m.). Levels on higher floor apartments were very similar to
those in lower floor apartments. As another example, there are plans to
build a high school over a very heavily used intersection in the Bronx. It
is not clear from press statements whether or not the potential pollution
impact has been taken into account in the design of the school.

Conclusion

Space limitations prohibit other detailed explorations of past applications of scientific information to citizen concerns. However, two general points can be made. The first is based on a trite principle of human
ecology, namely, that human beings are different from each other. In the
context of an air pollution discussion, this means that some people will
be more sensitive than others to a given amount of pollution. For example, people with pre-existing chronic bronchitis, emphysema, or asthma

uffer more when the sulfur oxide and particulate levels go up than people vithout these diseases. People who obtain CO from cigarette smoking may ;et an added dose, with even greater oxygen deprivation, if they venture nto, drive through, or work in an area high in CO. A person with heart lisease may be particularly sensitive to oxygen deprivation. In this re- ;ard, it should be pointed out that the laboratory studies done with CO vere done primarily on healthy men, often non-smokers, and not on the /ery young, the elderly, or people with chronic diseases. Just as the gen- ·ral environment is very complex, with many "micro-environments," so is he human species complex, with many subgroups. Care must be taken to)rotect the more sensitive groups from being unduly affected by a level)f air pollution or any other environmental insult, for that matter, that nay not affect less sensitive groups.

The second general point is exemplified by the apartments over the ιpproach to the George Washington Bridge, a specific set of buildings in ι specific place. We can see that there is reason to be concerned about he air pollution levels breathed by people living in these apartments, ιnd it is therefore important that such mistakes *not* be made in other pecific locations where such buildings may be constructed in the future. [t is not easy to eliminate these unnecessary environmental conditions ιfter the building is finished; this must occur before the construction of .he building—during design and planning. This is true for most environ- nental problems: they must be foreseen and dealt with as early as pos- ;ible, hopefully before they become serious.

What is pollution?

MARSHALL I. GOLDMAN AND ROBERT SHOOP

*Especially prepared for this study, "What Is Pollution?" is a quick sur-
vey of the various kinds of air and water pollutants, how they are
measured and how they are controlled. It is necessary to have a familiarity with the technical aspects of pollution before it is possible to
apply economic remedies. Robert Shoop of the Department of Biological Sciences, University of Rhode Island, collaborated with the
editor on this selection.*

As indicated earlier, pollution is a relative concept. Although almost
no substance exists in a pure state, it is only when the impurities rise
above a certain level that there is concern. Moreover the danger level
varies from case to case depending on the particular use for which the
water or air is intended. For example, water that is considered dangerous
for human use may be suitable for industrial use. Bethlehem Steel Company at Sparrows Point, Maryland, outside of Baltimore, uses the treated
sewage of Baltimore for making steel. It is cheaper than potable water
and does not affect the making of steel. More unusual, perhaps, water
that is considered safe for human beings may not be suitable for industrial purposes. Electric utilities using large quantities of water in their
boilers are forced to remove most of the minerals from the municipal
water they use. While the minerals are considered necessary and healthy
for human consumption, they have a corrosive effect on the insides of the
boilers.

Water Pollution

Classification. Because it is a relative concept and because there
are so many different elements which can have a detrimental effect on
water and air, many systems are used to classify the various types of pol-

ition. Let us consider water pollution first. Some biologists use the following classification for water pollution: (1) Putrescible materials; (2) Treated effluents; (3) Inert materials; (4) Toxic materials; (5) Radioactive elements and compounds.

1. Organic wastes subject to putrefaction are among the most common; one could almost say the "ordinary household variety." This includes materials like wastes from humans, paper pulp plants, and canneries. The goal in organic pollution control is to accelerate the process by which these materials are decomposed. In many cases this is done by adding bacteria to the already decaying material. When placed in a stream, such materials continue to decompose by consuming or absorbing large quantities of dissolved oxygen (DO_2) in the water. The greater the load of organic matter, the greater the demand on the dissolved oxygen and the more serious the pollution of the water. It is important to know therefore how heavy the concentration of organic matter is. One measure of pollution then is to calculate how much oxygen has been absorbed. This indicates the organic content of the water and is usually called the Biochemical Oxygen Demand (BOD) test. A BOD measurement indicates the amount of oxygen consumed at a temperature of 20 degrees centigrade over a five-day period. If too much oxygen is removed and it takes too long for it to be restored, there may be serious pollution. If the level of DO_2 is lowered enough, then putrefaction sets in and hydrogen sulphide gas is created. Even in mild cases of oxygen depletion, fish such as trout may be unable to find adequate oxygen for their needs. Consequently they may die or be forced to move elsewhere. Under such circumstances, scavenger fish like carp which require less oxygen are the only fish which can survive in such waters. Oxygen exhaustion is an especially serious problem for streams which receive waste from paper pulp mills. These plants may discharge unusually large quantities of effluent, depleting all available dissolved oxygen. Since paper pulp mills are normally located near woodlands they may be particularly destructive of natural wildlife (see Part IV).

Oxygen is restored to the water in a variety of ways. As it moves along, especially a fast moving stream, the water is aerated, which raises the dissolved oxygen content (DO_2) of the water. By photosynthesis, water plants release oxygen just as do land plants. However at night when the sun does not shine, no new oxygen is evolved; the amount of dissolved oxygen is usually lowest just before dawn. This indicates how there may be an interplay of air and water pollution. To the extent that air pollution creates smog and blocks out the sun, it takes longer to restore the oxygen content of the water.

Assuming that the stream is moving fast enough and that no new effluents are added to the water course, the stream should be able t cleanse itself of organic waste and restore dissolved oxygen by natura tumbling and aeration within a few miles of the source of pollution. His torically this was how wastes were purified. It was only when populatio became so dense that the process had to be supplemented with artificia methods. The organically polluted area of the stream can be delineate into three zones. The first zone is called the Zone of Immediate Pollutior Here the dissolved oxygen content is the lowest and the area is characte ized by an accumulation of the pollutant which affects the odor and colc of the water. The second zone is the Septic Zone where there may be lit tle dissolved oxygen and where some special animals with low oxyge requirements, such as certain snails, sewage worms, and rattail maggot may live. Because of the odor of decomposition, the area is usually eas to find. Finally there is the Zone of Recovery. Here the odors begin t disappear and fish like minnows and suckers begin to appear. There i usually a large bloom of plant life or algae on the border of Zones II an III. The increased quantity of oxygen combines with decomposed organ isms to break down the last traces of the organic material and creat highly fertile conditions. It should be remembered that the organic ma terial contains relatively large amounts of phosphates and nitrates, th prime components of some fertilizers.

Sewage in the water is potentially dangerous for human consump tion or swimming. However, it is not lack of oxygen *per se* or even th smell which makes it dangerous. Potential water contamination is meas ured by the presence of coliform bacteria, *Escherichia coli* (*E. coli*). I themselves coliform bacteria are not normally harmful, but since the accompany waste from human intestines, they indicate the presence o human waste which could contain various organisms, potentially th cause of dysentery, hepatitis, or typhoid.

One of the most serious examples of what can happen when a bod of water is taken for granted is Lake Erie. At one time it was a swim ming resort area and a fisherman's paradise. Now, some cities along Lak Erie require typhoid inoculations for anyone who intends to sail on th lake. The fish catch in 1956 of 6.9 million pounds of blue pike fell to 20 pounds in 1963. Lake Erie is dying. Scientists call the process eutrophica tion. Normally it should take hundreds of thousands of years. What hap pens ordinarily is that with time the outlet channels of most lakes begi to erode. As a result, more and more water is drained off from the lake As the lake becomes shallower, more of the lake bed is exposed to sun light and an algae growth begins to flourish. Dead particles from th algae gather together and form a sludge which collects along the shor with other algae growing there to form a swamp. At first this is good for the fish, because they now have more food and oxygen. But the dead algae are used as an energy source by the decomposer organisms and they in turn absorb the oxygen. As a result the fish begin to die or seek refuge elsewhere.

This process has been accelerated on Lake Erie with an almost unbelievable speed. Lake Erie presently is the shallowest of the Great Lakes. Moreover the sewage from homes and farms, which is rich in nitrogen and phosphorus, acts as fertilizer on the plant life of the lake. Consequently there are reports that a 2,600 square mile area in the middle of the Lake has been completely drained of oxygen. Even if various government and private organizations can be induced to cooperate in the cleaning of the lake, it is estimated that it will still cost billions of dollars to clean up. And then it is uncertain how much the lake can be revitalized.

2. Oxygen is most readily restored when the water is cool. The hotter it is, the lower the oxygen-holding capacity of the water. The bubbles that arise from heated water demonstrate what happens to the gases in hot water. Therefore the disposal of otherwise uncontaminated but hot water into a stream may be as harmful to the stream or lake as the introduction of organic matter. In either case, the oxygen content of the water is reduced. This helps to explain why water pollution is such a serious matter in tropical countries. The temperature is always so warm that it is difficult for the streams to absorb the necessary quantities of oxygen.

Hot water is put into water courses by industries that use water for cooling purposes. Steel mills, electric utilities, breweries and oil refineries use large quantities of water for cooling; in many cases the water is clean in all other respects. Nevertheless such water is usually very destructive to the fish population. In addition to lowering the oxygen content of the water, the higher temperature (thermal pollution) is harmful because many organisms live within a restricted temperature range and are unable to tolerate even moderate fluctuations in temperature.

3. Inert wastes are those which enter the water as solids but are not involved in chemical reactions. Such wastes include dust, metal filings, oil films, dust and silt from soil erosion. Unless removed by mechanical means such as filtering or allowing for sedimentation, these materials eventually settle to the bottom of the water course and block sunlight. As a result, plant life is affected which in turn cuts off the food supply of the fish and other animal populations. For example, the oyster beds off the coast of Connecticut, Rhode Island, and Massachusetts have been buried with inert wastes. Pollution from inert wastes is also a serious problem in areas located near strip mines. The removal of the restraining

vegetation and the exposure of the unprotected earth results in soil ero sion. The soil is invariably washed away as silt to nearby water course With the topsoil gone, there is nothing left to hold back the mine acid wastes from washing into the rivers and streams.

4. Toxic wastes are those which do not easily settle out and are no easily broken down by biological means. They also tend to be poisonou when consumed or contacted by plants and animals. Pickling liquo which is a by-product of steelmaking, and phenols from coke ovens ar examples of toxic materials. Pesticides and herbicides which wash off th land into the sewers are other examples. Water courses are normally ur able to neutralize such toxic compounds as they do organic wastes. Co sequently special treatment facilities involving the addition of chemica neutralizers are required.

Mercury is an example of a toxic waste that has aroused widesprea concern recently. However, even normally harmless inorganic substance may constitute a threat if they are discharged into the water in exces amounts. For instance, with the increasing use of salt on the highway to melt snow, some communities like Bedford, Massachusetts have re ported that the chloride content of their water supplies has risen fro 8 parts per million (ppm) in 1955 to 290 ppm in the summer of 197(The maximum allowable limit as set by the federal government is 25 ppm. Since such high levels of chloride in the water are potentially ha ardous for people with high blood pressure, cirrhosis of the liver, an heart congestion, Bedford has prohibited the use of salt on its streets.

5. Radioactive substances are even more difficult to handle. Thes materials are produced in the processing of uranium and other radio active substances or in testing of the thermonuclear devices that produc nuclides in blast devices and fallout. It may take years for the level c radioactivity in the water to fall. Practically the only way known to di: pose of such materials is to dump them into the ocean beyond the cor tinental shelf or to pump them into abandoned mines and deep well. Even so there is fear that this may still cause pollution of oceans or ur derground water supplies.

Measurement.   It is not always easy to measure water pollution. I extreme cases, our noses are some help. In other cases, the sudden ar pearance of thousands of dead fish indicates that something is wron{ However, with all the new compounds being used, there are many po sonous mixtures we are unable to recognize and many that we as ye have not bothered to measure because we have not recognized the potential danger. After large numbers of dead fish were washed ashor on the banks of the Mississippi River, investigators discovered that m ute quantities of the chemical compounds endrin and dieldrin had been released into the river. Almost impossible to measure and trace, many scientists were convinced that one of those two pesticides has been the source of the trouble.

In order to assign responsibility, both moral and financial, to those who pollute, it is necessary to develop accurate and inexpensive measuring methods. In some cases the cost of measurement may exceed the economies to be gained by cleaning up the pollution. Ideally what is needed is an inexpensive device that without supervision can measure changes in pollution caused by a variety of pollutants. At the present time, scientists use physical methods for measuring pollution, such as inserting meters which measure the electrical conductivity of the water. This indicates the level of dissolved salts. Perhaps the most common test is to measure for the BOD level of the water. Biological methods are also used. For example a group of fish may be put into various dilutions of an effluent to see if they die, and if so, how rapidly.

Treatment.   Except for unusual kinds of industrial waste, sewage treatment is quite standardized and not especially sophisticated. There are three stages and some municipalities use all three. Some use none; most use two. Primary treatment consists of mechanical treatment such as screening and gravity settling. Relatively simple and inexpensive, primary treatment removes about 60–80 percent of the sedimentation, about one-third of the BOD and a little nitrogen and phosphorus. Secondary treatment involves subjecting the screened waste of the primary stage to a controlled form of oxidation using biological and bacterial action. A well-run plant can remove 80–90 percent of the BOD and about one-third of the phosphorus and one-half of the nitrogen. The biological treatment is carried out using either the biological filter or activated sludge process. The system of biological filters involves taking the sewage from the sedimentation tanks and passing it over something like a bed of rocks through which air is circulated. Bacteria form on the rocks and interact and oxidize the sewage as it moves by. In the activated sludge system, compressed air is forced through tanks containing sewage and activated sludge which also interacts and digests and oxidizes the impurities. Both methods are efforts to duplicate the natural processes of rivers and streams.

Unfortunately biological treatment is not very effective in removing the phosphates and nitrates. Tertiary treatment attempts to make up for this and also complete the removal of up to 99 percent of the BOD. One way to do this involves extensive aeration of the water or filtration over natural rocks and sand and in some experiments, charcoal. In some lim-

ited instances, the water is then returned to the drinking water supply ;
in Windhoek, South Africa and Lubbock, Texas.

In a large number of cases, after secondary treatment, the effluen
is discharged into a water course where hopefully, if the river is healtl
and no other pollutants are immediately added, it will be purified natu
rally. Most of us do not realize it, but just as Tom Lehrer says in his son
on p. vii, it often happens that the water we drink is taken from a wate
course used by the upstream residents as a sewer. Before the water
used for drinking, however, it is subjected to chlorination. This is usual
done even where water is obtained from subsurface wells. As indicativ
of how new (and relatively primitive) the whole process of water purifica
tion is, the purifying effect of chlorine was only discovered in 1908.

The existence of a secondary or even tertiary treatment plant doe
not guarantee pollution control. As we have seen, in some cases industri;
waste requires special processing which may not be provided in a muni
ipal facility. It is therefore necessary for the industry to treat its own sev
age or see that the municipal facilities provide the necessary care. I
other cases, the sewage treatment system may have a limited capacit
and is therefore unable to handle suddenly enlarged loads. Most Amer
can sewer systems are quite old. As a result they contain only one set o
sewer pipes that are made to handle both household sewage and storn
drainage. Normally this is adequate, but during a rain storm or hurrican·
the sewer network and the sewage treatment plant lack the capacity t
process the sudden extra load. Accordingly it is usually necessary t
divert the combined load of rain water and untreated sewage directl
into the river or other nearby body of water. One cure for this is to buil
another set of sewer pipes just for drainage of rainwater which tradi
tionally has required less treatment. Now however some sanitary eng
neers are no longer sure that this is wise. Increasingly, storm water carrie
with it contaminating chemicals such as snow melting salt, pesticides an
herbicides as well as oil and asphalt from automobiles and highway
Another solution is to build storage lagoons which can be used to hol
industrial and storm wastes until such time as the storm has passed an
the sewage treatment plant is able to resume normal operation.

Air Pollution

Classification.   Essentially the nature of air and water pollution
the same. In each case the impurities that are introduced are not al
sorbed adequately or rapidly enough. It is necessary to remove or dilut
the more disruptive elements and gases from the exhaust smoke befor
they are released into the atmosphere. The problem is more serious tha

often appears. Although we tend to behave as if the atmosphere were infinite in depth, scientists tell us that it is actually quite thin. Consequently there is not that much air around after all with which to dilute smoke and gaseous waste.

Among the usual wastes released into the air are lead and carbon dioxide, sulfur dioxide, carbon monoxide, nitrogen oxide and suspended particulates like dust. Lead is commonly used in gasoline because of its antiknock properties. It is considered very hazardous, but not much is known about how much the human system can absorb from the lead generated by the burning of gasoline. Also not much is known about the hazards which exist from increasing the amount of carbon dioxide in the atmosphere. There is some fear among the more pessimistic scientists that the earth's natural balance has already been thrown off. Among other effects, this could cause an increase in the earth's temperature of about one degree which could result in the advance of deserts and the inundation of coastal areas as icebergs begin to melt. Other scientists however say we may be faced with exactly the opposite kind of threat. As air pollution blocks out the sun, the earth's temperature may drop causing the ice cap to spread. One way or the other, the disaster buffs should be delighted.

Sulfur dioxide is produced by the burning of coal and oil fuels, depending on amounts of sulfur contained in the particular fuel. Bituminous coal and residual or heavy oil from Venezuela contain large quantities of sulfur. Unfortunately, these are also the cheapest fuels and therefore they are normally much in demand. Coal provides over 50 percent of the nation's electric power. Sulfur dioxide is usually a problem in areas where electricity is generated in coal-fueled furnaces, where coal is used as a source of home fuel and where there are large steel mills and other coal-burning industries.

Carbon monoxide and nitrogen oxide are especially troublesome in large metropolitan areas where there are large concentrations of automobiles. Of course, carbon monoxide is poisonous and is often the cause of death, especially in the winter when drivers and parkers are more likely to keep windows closed and motors running. Even the normal bystander may be affected. Carbon monoxide is considered fatal when there are over 400 parts per million (ppm) in the air. Readings of 15 ppm were recorded in the open street at Columbus Circle in New York City in 1970. Chicago has had readings of 39 ppm, and on Los Angeles freeways levels of 54 ppm have been recorded in slow moving traffic. Automobile fumes are so concentrated in Tokyo that visiting Americans and some Japanese suffer from drowsiness that is called "Tokyo asthma." Nitrogen oxide, which no one yet knows how to eliminate cheaply from automobile ex-

haust, is one of the main causes of automobile-generated smog. Recentl
the increased use of jet airplanes has intensified the problem. There ar
estimates indicating that the fumes from one jet plane throw off pollutio
equivalent to 6,800 cars.

When nitrogen oxides and hydrocarbons are exposed to the ultra
violet rays of the sun, a photochemical reaction takes place which create
oxidants such as ozone. At one time everyone thought that ozone was
beneficial substance and names like Ozone Park were designated fo
towns to indicate fresh air and purity. Now we know that when th
amount of oxidants in the air reads 0.15 ppm for a period of one hou
there will be plant damage and discomfort. In the first half of 1964, read
ings as high as 0.25 ppm were recorded in Cincinnati, Philadelphia, S
Louis, Los Angeles and Washington, D.C. Ozone has greater oxidizin
power than oxygen, and ozone damage in such forms as rust was est
mated to total about $25 million a year in twenty-two states.

Particulate matter or plain old dust and dirt is the form of air po
lution that is most familiar to the average citizen. Few of us have ev
choked on sulfur dioxide. While we may have been caught in the fume
of a bus or pleased by the smell of burning leaves in autumn, we ar
most conscious of the dust and dirt which descend on us, especially i
large cities. Some of this dirt is also a by-product of the combustion c
electric utilities and home furnaces, but it is primarily caused by th
smoke from incinerators and dumps. To eliminate the more seriou
sources of such pollution, some cities have prohibited the use of ope
fires and have demanded the use of specially designed incinerators.

The amount and type of pollution enjoyed by any given America
city is dependent on its topography, industrial makeup and legal injun
tions. New York City has a very high level of sulfur dioxide and su
pended particulate matter. This is so despite the fact that New York Cit
is a city without much heavy industry and that it is well-situated: th
prevailing winds are not impeded as they blow the polluted air into th
sea. However part of New York City's problem is due to the fact that b
fore the winds blow out to sea, they come in from highly industrialize
districts of New Jersey. This is fine for New Jersey which thereby solv
much of its air pollution problem by exporting it to New York City. It
estimated that one third of New York City's pollution is generated in Ne
Jersey. This illustrates the need for regional rather than unilateral contr
of pollution.

It is also worth noting that one of the most heavily polluted are
in New York City is the high-rent district of the upper East Side. Th
luxurious apartments just north of the United Nations register one
the highest densities of sulfur dioxide of any place in the whole countr

The reason is that the New York electric utility company, Consolidated Edison, has located several large and not always efficient electric generating stations along the East River. As the apartment houses grow in height, their terraces and windows open up directly on to the stacks of these power plants.

Los Angeles, which we think of as having the more serious air pollution problem, actually has very low readings of sulfur dioxide and suspended particulate matter, but it has the highest amount of nitrogen oxide. As their smog problem intensified, inhabitants of Los Angeles virtually banned all coal fires and all industrial smoke and started to bury their trash rather than burn it. But the nitrogen oxide produced by their three million automobiles has more than offset the improvements elsewhere and the smog problem continues to burden the city. To control it, Los Angeles enacted a law requiring the installation of combustion control devices on all new automobiles produced in 1966 and thereafter. A similar law in somewhat less stringent form was subsequently passed at the federal level effective with the 1968 model car; it is questionable if such a law will be enough. Each year the problem becomes more serious, largely because the number of automobiles on the streets increases so rapidly. In recognition of this, the Environmental Protection Agency set some rigorous requirements for 1975 that if enforced may require some major cities like New York, Chicago, St. Louis, Baltimore, Hartford, Buffalo and Philadelphia to ban or restrict traffic at certain times during the day. By 1975, the standard for sulfur oxides is to be lowered to 80 micrograms per cubic meter (.03 ppm), with a maximum three-hour concentration never to exceed 1,300 micrograms per cubic meter (.05 ppm); for hydrocarbons the limit for a three-hour period is to be 160 micrograms per cubic meter (.24 ppm); for nitrogen oxides the limit will be 100 micrograms per cubic meter (.05 ppm) and for carbon monoxide, the maximum one-hour concentration is to be 160 micrograms per cubic meter (.08 ppm). For most of the cities mentioned, this would necessitate a reduction of at least 50 percent from the levels that prevailed in 1970. For the automobile makers, attainment of these standards will require a 90 percent drop in existing emission levels.

To the surprise of some, the increasing number of small foreign cars actually intensifies the situation. For a long time it was assumed that small cars like the Volkswagen emitted fewer gases than larger American-made vehicles. In tests conducted by the California Air Resources Board in 1970 and printed in *Environment* in September 1970, the Volkswagen was found to emit more hydrocarbons and carbon monoxide per mile than any of the three major American automobile manufacturers. Only in the emission of nitrogen oxide was it a bit cleaner. Other foreign cars such as the Toyota, Opel, and Mercedes were no better and often much worse.

On occasion the day-to-day hazards of air pollution are aggravated by the meteorological phenomenon called air inversion. In simple terms, this means that normal air conditions are inverted or just the opposite of what they should normally be. What usually happens to warm air in the home? It rises. The same principle applies in the city. Warm factory smoke normally rises and takes with it the industrial waste where it is dissipated in the upper atmosphere. But occasionally on a clear night there may be a sudden loss of heat at ground level. Therefore early the next morning before the sun comes out, it may be warmer in the upper air than in the air at ground level, the reverse of what it should be. If there is a lack of wind, this can lead to an inversion. The cool air immediately above the ground will not rise. Instead it forms a blanket and prevents the new warm air from the factories from rising. As the sun begins to warm up the atmosphere, the cold air is usually baked enough to dissipate the inversion layer by midmorning, but occasionally it may last for several days. It is the inversion layer holding the smoke and gas from industry and automobile exhaust which explains the early morning blue haze in most large American cities. Inversions also take place in fogs preventing smoke and other contaminants from rising.

Cities surrounded by hills, like Los Angeles, are doubly troubled. On the one hand, the hills block the wind which might clear the air so that the city's air forms a lid over the city. On the other hand, the cold air from the hills often slips down at night, so that the cold air blocks the rise of the next day's wastes. Therefore cities in valleys or alongside mountainous areas are especially susceptible to air inversion and must be extremely cautious about the kind of industry they allow in the area. This is an ever recurring problem since industries are usually attracted to valleys because their streams make convenient receptacles for water waste. But the conditions that make possible good water disposal make bad air disposal. Thus the river valleys of West Virginia, with their chemical, coal and iron industries, are considered to suffer more from air pollution than any other region in the country.

Treatment.   The technology of smoke prevention is also relatively primitive. Essentially there are four approaches, most of which are used simultaneously.

1. The best solution is to ensure that there is good combustion. The combustion chamber must be well-supplied with oxygen and a good draft so that the temperature of the fire is as hot as possible.

This will eliminate much of the dark smoke which contains incompletely burned dust and ashes. The importance of good combustion for smoke control can be easily demonstrated. Light a cigarette. Now take a lighted match and place the flame directly in the smoke rising from the cigarette and notice what happens to the smoke. The same principle applies to an incinerator as well as to an automobile. This is why open fires are banned at dumps and junkyards. Pollution control devices on cars are basically an attempt to do nothing more than ensure a more complete combustion of the car's gasoline. With a similar aim in mind, pollution control officials have discovered that automobile-created pollution can be reduced by accelerating the flow of traffic. The faster the engine runs, the better the combustion. Similarly the faster the car moves, the more its fumes are disbursed over a wider area. Automobile pollution is at its worst when a vehicle is stationary with its engine running or when a car is decelerating.

2. The second technique is to use some kind of mechanical device. This may be by means of filtration or the use of a screen, or simply a gravity process whereby particulate matter is passed over several settling chambers before being allowed to reach the chimney flue. This process may be further supplemented by the use of electrostatic precipitators costing as much as half a million dollars. These units consist of electrically charged wires that are run through a series of flues on the way to the chimney stacks. They attract much of the ash and dust that would otherwise fly into the open air. Because high voltage does not have much effect on oil, electrostatic precipitators seem to be useful only with coal fires. As more and more electric utility companies switch to oil fires to reduce sulfur dioxide release into the air, the electrostatic precipitator is of no help. Of course another mechanical solution is to build the chimney stacks higher so that the fumes will be diffused at a higher level. Often this is necessary just to prevent the fumes from blowing directly into the windows of new office and apartment buildings. We saw how this affected the apartments on New York City's upper East Side. In hopes that it will alleviate some of the trouble, Consolidated Edison is planning to spend $14 million on installing new boilers and eliminating eight of their Manhattan stacks. The present eight stacks, which are 270 feet high, will be replaced with two stacks which will be 509 feet, or about fifty stories high.

3. The most effective smoke control is obtained by the use of either water or oil scrubbers. As indicated, most of the ash and even some of the gases are washed and made reasonably clean. Scrubbers however can be even more expensive than the electrostatic precipitators and are not in common use. Moreover while scrubbers usually reduce chimney omis-

sions to a white plume of steam, there is merely a reshuffling of the problem because the liquid used for scrubbing is now polluted and must be treated before it is released into a public body of water.

4. Finally, some factory fumes must be subjected to chemical treatment. Only by neutralizing them or forming some harmless and on occasion valuable by-product can the smoke in such cases be rendered unobjectionable. It is also possible to extract undesirable elements before combustion. Oil officials estimate that it would cost from 40 to 70 cents a barrel to reduce the sulfur content of oil from 2.8 percent to the much cleaner 1 percent that was required for a time in large cities such as New York City and Boston.

PART 2

ECONOMIC ANALYSIS

The role of government
in a free society

MILTON FRIEDMAN

Milton Friedman is Paul Snowden Russell Professor of Economics at the University of Chicago. He is one of the most prominent defenders of the sanctity of the private market economy. In this essay he tries to show how the unregulated market system in almost all instances benefits the public. He points out, however, that certain private actions, among them private pollution, involve involuntary exchange and thereby deny appropriate market compensation to the affected individuals.

A common objection to totalitarian societies is that they regard the end as justifying the means. Taken literally, this objection is clearly illogical. If the end does not justify the means, what does? But this easy answer does not dispose of the objection; it simply shows that the objection is not well put. To deny that the end justifies the means is indirectly to assert that the end in question is not the ultimate end, that the ultimate end is itself the use of the proper means. Desirable or not, any end that can be attained only by the use of bad means must give way to the more basic end of the use of acceptable means.

To the liberal,* the appropriate means are free discussion and voluntary cooperation, which implies that any form of coercion is inappropriate. The ideal is unanimity among responsible individuals achieved on the basis of free and full discussion. This is another way of expressing the goal of freedom.

Reprinted with permission from *Capitalism and Freedom,* © 1962 by the University of Chicago Press.
*Friedman here uses the word "liberal" as it was used in nineteenth century England. Then a liberal was one who believed in the limitation rather than the expansion of governmental activity in business affairs.

From this standpoint, the role of the market, as already noted, is that it permits unanimity without conformity; that it is a system of effec- tively proportional representation. On the other hand, the characteristic feature of action through explicitly political channels is that it tends to require or to enforce substantial conformity. The typical issue must be decided "yes" or "no"; at most, provision can be made for a fairly limited number of alternatives. Even the use of proportional representation in its explicitly political form does not alter this conclusion. The number of separate groups that can in fact be represented is narrowly limited, enor- mously so by comparison with the proportional representation of the market. More important, the fact that the final outcome generally must be a law applicable to all groups, rather than separate legislative enact- ments for each "party" represented, means that proportional representa- tion in its political version, far from permitting unanimity without con- formity, tends toward ineffectiveness and fragmentation. It thereby operates to destroy any consensus on which unanimity with conformity can rest.

There are clearly some matters with respect to which effective pro- portional representation is impossible. I cannot get the amount of na- tional defense I want and you, a different amount. With respect to such indivisible matters we can discuss, and argue and vote. But having de- cided, we must conform. It is precisely the existence of such indivisible matters—protection of the individual and the nation from coercion is clearly the most basic—that prevents exclusive reliance on individual ac- tion through the market. If we are to use some of our resources for such indivisible items, we must employ political channels to reconcile differ- ences.

The use of political channels, while inevitable, tends to strain the social cohesion essential for a stable society. The strain is least if agree- ment for joint action need be reached only on a limited range of issues on which people in any event have common views. Every extension of the range of issues for which explicit agreement is sought strains further the delicate threads that hold society together. If it goes so far as to touch an issue on which men feel deeply yet differently, it may well disrupt the society. Fundamental differences in basic values can seldom if ever be re- solved at the ballot box; ultimately they can only be decided, though not resolved, by conflict. The religious and civil wars of history are a bloody testament to this judgment.

The widespread use of the market reduces the strain on the social fabric by rendering conformity unnecessary with respect to any activities it encompasses. The wider the range of activities covered by the market the fewer are the issues on which explicitly political decisions are required and hence on which it is necessary to achieve agreement. In turn, the

ewer the issues on which agreement is necessary, the greater is the likelihood of getting agreement while maintaining a free society.

Unanimity is, of course, an ideal. In practice, we can afford neither
the time nor the effort that would be required to achieve complete unanimity on every issue. We must perforce accept something less. We are
thus led to accept majority rule in one form or another as an expedient.
That majority rule is an expedient rather than itself a basic principle is
clearly shown by the fact that our willingness to resort to majority rule,
and the size of the majority we require, themselves depend on the seriousness of the issue involved. If the matter is of little moment and the
minority has no strong feelings about being overruled, a bare plurality
will suffice. On the other hand, if the minority feels strongly about the
issue involved, even a bare majority will not do. Few of us would be
willing to have issues of free speech, for example, decided by a bare
majority. Our legal structure is full of such distinctions among kinds of
issues that require different kinds of majorities. At the extreme are those
issues embodied in the Constitution. These are the principles that are so
important that we are willing to make minimal concessions to expediency.
Something like essential consensus was achieved initially in accepting
them, and we require something like essential consensus for a change in
them.

The self-denying ordinance to refrain from majority rule on certain
kinds of issues that is embodied in our Constitution and in similar written
constitutions elsewhere, and the specific provisions in these constitutions
or their equivalents prohibiting coercion of individuals, are themselves to
be regarded as reached by free discussion and as reflecting essential unanimity about means.

I turn now to consider more specifically, though still in very broad
terms, what the areas are that cannot be handled through the market at
all, or can be handled only at so great a cost that the use of political
channels may be preferable.

GOVERNMENT AS RULE-MAKER AND UMPIRE

It is important to distinguish the day-to-day activities of people from the
general customary and legal framework within which these take place.
The day-to-day activities are like the actions of the participants in a game
when they are playing it; the framework, like the rules of the game they
play. And just as a good game requires acceptance by the players both
of the rules and of the umpire to interpret and enforce them, so a good
society requires that its members agree on the general conditions that
will govern relations among them, on some means of arbitrating different

interpretations of these conditions, and on some device for enforcing compliance with the generally accepted rules. As in games, so also in society, most of the general conditions are the unintended outcome of custom, accepted unthinkingly. At most, we consider explicitly only minor modifications in them, though the cumulative effect of a series of minor modifications may be a drastic alteration in the character of the game or of the society. In both games and society also, no set of rules can prevail unless most participants most of the time conform to them without external sanctions; unless that is, there is a broad underlying social consensus. But we cannot rely on custom or on this consensus alone to interpret and to enforce the rules; we need an umpire. These then are the basic roles of government in a free society: to provide a means whereby we can modify the rules, to mediate differences among us on the meaning of the rules, and to enforce compliance with the rules on the part of those few who would otherwise not play the game.

The need for government in these respects arises because absolute freedom is impossible. However attractive anarchy may be as a philosophy, it is not feasible in a world of imperfect men. Men's freedoms can conflict, and when they do, one man's freedom must be limited to preserve another's—as a Supreme Court Justice once put it, "My freedom to move my fist must be limited by the proximity of your chin."

The major problem in deciding the appropriate activities of government is how to resolve such conflicts among the freedoms of different individuals. In some cases, the answer is easy. There is little difficulty in attaining near unanimity to the proposition that one man's freedom to murder his neighbor must be sacrificed to preserve the freedom of the other man to live. In other cases, the answer is difficult. In the economic area, a major problem arises in respect of the conflict between freedom to combine and freedom to compete. What meaning is to be attributed to "free" as modifying "enterprise"? In the United States, "free" has been understood to mean that anyone is free to set up an enterprise, which means that existing enterprises are not free to keep out competitors except by selling a better product at the same price or the same product at a lower price. In the continental tradition, on the other hand, the meaning has generally been that enterprises are free to do what they want, including the fixing of prices, division of markets, and the adoption of other techniques to keep out potential competitors. Perhaps the most difficult specific problem in this area arises with respect to combinations among laborers, where the problem of freedom to combine and freedom to compete is particularly acute.

A still more basic economic area in which the answer is both difficult and important is the definition of property rights. The notion of property, as it has developed over centuries and as it is embodied in our egal codes, has become so much a part of us that we tend to take it for granted, and fail to recognize the extent to which just what constitutes property and what rights the ownership of property confers are complex social creations rather than self-evident propositions. Does my having title to land, for example, and my freedom to use my property as I wish, permit me to deny to someone else the right to fly over my land in his airplane? Or does his right to use his airplane take precedence? Or does this depend on how high he flies? Or how much noise he makes? Does voluntary exchange require that he pay me for the privilege of flying over my land? Or that I must pay him to refrain from flying over it? The mere mention of royalties, copyrights, patents, shares of stock in corporations, riparian rights and the like may perhaps emphasize the role of generally accepted social rules in the very definition of property. It may suggest also that, in many cases, the existence of a well specified and generally accepted definition of property is far more important than just what the definition is.

Another economic area that raises particularly difficult problems is the monetary system. Government responsibility for the monetary system has long been recognized. It is explicitly provided for in the constitutional provision which gives Congress the power "to coin money, regulate the value thereof, and of foreign coin." There is probably no other area of economic activity with respect to which government action has been so uniformly accepted. This habitual and by now almost unthinking acceptance of governmental responsibility makes thorough understanding of the grounds for such responsibility all the more necessary, since it enhances the danger that the scope of government will spread from activities that are, to those that are not, appropriate in a free society, from providing a monetary framework to determining the allocation of resources among individuals.

In summary, the organization of economic activity through voluntary exchange presumes that we have provided, through government, for the maintenance of law and order to prevent coercion of one individual by another, the enforcement of contracts voluntarily entered into, the definition of the meaning of property rights, the interpretation and enforcement of such rights, and the provision of a monetary framework.

ACTION THROUGH GOVERNMENT ON GROUNDS OF TECHNICAL MONOPOLY AND NEIGHBORHOOD EFFECTS

The role of government just considered is to do something that the market cannot do for itself, namely, to determine, arbitrate, and enforce the rules of the game. We may also want to do through government some things that might conceivably be done through the market but that tech nical or similar conditions render it difficult to do in that way. These a reduce to cases in which strictly voluntary exchange is either exceedingl costly or practically impossible. There are two general classes of suc cases: monopoly and similar market imperfections, and neighborhoo effects.

Exchange is truly voluntary only when nearly equivalent alterna tives exist. Monopoly implies the absence of alternatives and thereby ir hibits effective freedom of exchange. In practice, monopoly frequently, i not generally, arises from government support or from collusive agree ments among individuals. With respect to these, the problem is either t avoid governmental fostering of monopoly or to stimulate the effectiv enforcement of rules such as those embodied in our antitrust laws. How ever, monopoly may also arise because it is technically efficient to hav a single producer or enterprise. I venture to suggest that such cases ar more limited than is supposed but they unquestionably do arise. A simpl example is perhaps the provision of telephone services within a commu nity. I shall refer to such cases as "technical" monopoly.

When technical conditions make a monopoly the natural outcom of competitive market forces, there are only three alternatives that seen available: private monopoly, public monopoly, or public regulation. Al three are bad so we must choose among evils. Henry Simons, observin public regulation of monopoly in the United States, found the results s distasteful that he concluded public monopoly would be a lesser evil Walter Eucken, a noted German liberal, observing public monopoly in German railroads, found the results so distasteful that he concluded pub lic regulation would be a lesser evil. Having learned from both, I reluc tantly conclude that, if tolerable, private monopoly may be the least o the evils.

If society were static so that the conditions which give rise to a tech nical monopoly were sure to remain, I would have little confidence ir this solution. In a rapidly changing society, however, the conditions mak ing for technical monopoly frequently change, and I suspect that botl public regulation and public monopoly are likely to be less responsive t such changes in conditions, to be less readily capable of elimination, thar private monopoly.

Railroads in the United States are an excellent example. A large degree of monopoly in railroads was perhaps inevitable on technica grounds in the nineteenth century. This was the justification for the In terstate Commerce Commission. But conditions have changed. The emer gence of road and air transport has reduced the monopoly element ir railroads to negligible proportions. Yet we have not eliminated the ICC

On the contrary, the ICC, which started out as an agency to protect the public from exploitation by the railroads, has become an agency to protect railroads from competition by trucks and other means of transport, and more recently even to protect existing truck companies from competition by new entrants.

Technical monopoly may on occasion justify a *de facto* public monopoly. It cannot by itself justify a public monopoly achieved by making it illegal for anyone else to compete. For example, there is no way to justify our present public monopoly of the post office. It may be argued that the carrying of mail is a technical monopoly and that a government monopoly is the least of evils. Along these lines, one could perhaps justify a government post office but not the present law, which makes it illegal for anybody else to carry mail. If the delivery of mail is a technical monopoly, no one will be able to succeed in competition with the government. If it is not, there is no reason why the government should be engaged in it. The only way to find out is to leave other people free to enter.

A second general class of cases in which strictly voluntary exchange is impossible arises when actions of individuals have effects on other individuals for which it is not feasible to charge or recompense them. This is the problem of "neighborhood effects". An obvious example is the pollution of a stream. The man who pollutes a stream is in effect forcing others to exchange good water for bad. These others might be willing to make the exchange at a price. But it is not feasible for them, acting individually, to avoid the exchange or to enforce appropriate compensation.

Parks are an interesting example because they illustrate the difference between cases that can and cases that cannot be justified by neighborhood effects, and because almost everyone at first sight regards the conduct of national parks as obviously a valid function of government. In fact, however, neighborhood effects may justify a city park, they do not justify a national park, like Yellowstone National Park or the Grand Canyon. What is the fundamental difference between the two? For the city park, it is extremely difficult to identify the people who benefit from it and to charge them for the benefits which they receive. If there is a park in the middle of the city, the houses on all sides get the benefit of the open space, and people who walk through it or by it also benefit. To maintain toll collectors at the gates or to impose annual charges per window overlooking the park would be very expensive and difficult. The entrances to a national park like Yellowstone, on the other hand, are few; most of the people who come stay for a considerable period of time and it is perfectly feasible to set up toll gates and collect admission charges. This is indeed now done, though the charges do not cover the whole costs. If the public wants this kind of an activity enough to pay for it, private enterprises will have every incentive to provide such parks. And, of course, there are many private enterprises of this nature now in existence. I cannot myself conjure up any neighborhood effects or important monopoly effects that would justify governmental activity in this area.

Considerations like those I have treated under the heading of neighborhood effects have been used to rationalize almost every conceivable intervention. In many instances, however, this rationalization is special pleading rather than a legitimate application of the concept of neighborhood effects. Neighborhood effects cut both ways. They can be a reason for limiting the activities of government as well as for expanding them. Neighborhood effects impede voluntary exchange because it is difficult to identify the effects on third parties and to measure their magnitude; but this difficulty is present in governmental activity as well. It is hard to know when neighborhood effects are sufficiently large to justify particular costs in overcoming them and even harder to distribute the costs in an appropriate fashion. Consequently, when government engages in activities to overcome neighborhood effects, it will in part introduce an additional set of neighborhood effects by failing to charge or to compensate individuals properly. Whether the original or the new neighborhood effects are the more serious can only be judged by the facts of the individual case, and even then, only very approximately. Furthermore, the use of government to overcome neighborhood effects itself has an extremely important neighborhood effect which is unrelated to the particular occasion for government action. Every act of government intervention limits the area of individual freedom directly and threatens the preservation of freedom indirectly.

Social costs of business enterprise

KARL WILLIAM KAPP

Karl William Kapp, a noted social economist, discusses the reasons why social costs must constitute a special category in economic analysis. He also describes some of the specific costs to society which are caused by water and air pollution. Of special interest is his theory that damages to society from an activity like pollution do not tend to level off or decrease naturally. Rather such damages contribute to one another without bringing about any natural economic counterbalancing force to check their cumulative growth.

THE NATURE AND SIGNIFICANCE OF SOCIAL COSTS

In order to be recognized as social costs, harmful effects and inefficiences must have two characteristics. It must be possible to avoid them and they must be part of the course of productive activities and be shifted to third persons or the community at large. For instance, pollution of the environment by various types of contaminants can be traced to productive activities and can be shown to be man-made and avoidable.

The basic causes of social costs are to be found in the fact that the pursuit of private gain places a premium on the minimization of the private costs of current production. Therefore, the greater the reliance on private incentives, the greater the probability of social costs. The more reliance an economic system places on private incentives and the pursuit of private gain, the greater the danger that it will give rise to external "unpaid" social costs unless appropriate measures are taken to avoid or at least minimize these costs. From this it would follow that a decentral-

From *Social Costs of Business Enterprise,* © K. William Kapp, 1963. Asia Publishing House, New York. References to statistical investigations and estimates of losses have been omitted.

ized planned economy which makes extensive use of private incentive such as bonuses to its managers in order to assure the attainment of it targets and objectives, will hardly be immune to social costs. Evidence o certain inefficiencies in the Soviet economy supports this conclusion. In s far as social costs are the result of the minimization of the internal cost of the firm, it is possible to regard the whole process as guidance of redistribution of income. By shifting part of the costs of production t third persons or to the community at large, producers are able to appro priate a larger share of the natural product than they would otherwise b able to do. Alternatively, it may be claimed that consumers who pur chase the products will get them at lower prices than they would hav been able to do had producers been forced to pay the total costs of pro duction. And similarly, the enforcement of preventive legislation has a re distributive effect. There is thus a problem of the incidence of social cost and of the costs and benefits of measures designed to prevent the socia losses caused by private productive activities. How large a share of th national income is thus redistributed is a problem which cannot concer us here. It is needless to add that the fact that problems of social cost raise issues of income redistribution makes them matters of political con troversy and political power.

The main body of neoclassical value theory has continued to regarc social losses as accidental and exceptional cases or as minor disturbances To dismiss the entire problem of social costs in this manner begs the question. For whether or not these costs are isolated cases and minor dis turbances can be decided only after their significance and their probable magnitude have been explored. Similarly, to dismiss the problem of social costs on the ground that in some instance remedial measures have beer taken by governments and private organizations misses the important question as to whether such measures are adequate and effective. This question too can be answered only after an attempt has been made to comprehend the nature and possible magnitude of the social cost of pro duction. Close analysis of existing preventive regulations reveals that present restraints are still highly ineffective in minimizing social losses, and that in many instances social costs have found either no or only the most rudimentary recognition.

It has to be admitted that some social costs are highly complex and composite in character and can be evaluated only in terms of the im portance which organized society attributes to both the tangible and the intangible values involved. Such social evaluations do not, however, pose an entirely *new* problem. The formulation of public policy nearly always requires an evaluation of "means" and "ends" whose relative importance can only be estimated due to the fact that a substantial proportion of all costs" and "returns" of economic policies are "political" and intangible in character. Of course, this is not to say that policy formulation on the basis of general estimates of possible costs and returns has been fully explored and offers no further theoretical difficulties. Quite on the contrary, the issues raised by the concept of "social value" and "social evaluation" belong to the most important unsolved problems of economic science.

The social scientist must resist the temptation either to ignore social costs because they require an evaluation by society or to introduce his own standards and preferences into the discussion of their evaluation. For, to do the former and to leave social losses out of account because they are "external" and "noneconomic" in character, would be equivalent to attributing no or "zero" value to all social damages which is no less arbitrary and subjective a judgment than any positive or negative evaluation of social costs. For the social scientist to evaluate social costs in terms of his own standards and preferences would mean the introduction of highly subjective value judgments into economic analysis and would make the generalization derived therefrom equally subjective, arbitrary and problematical. Here, as in all matters of social evaluation, we would be on safer grounds if we could rely on objective standards of social minima and measure social costs in terms of shortfalls or deficiencies from such minima.

CUMULATIVE CAUSATION AND SOCIAL COSTS

One of the central tasks of the theory of social costs and indeed the key objective of the present study is to trace the causal relationships between various productive activities and business practices on the one hand and of significant social losses and damages on the other. This undertaking will be rendered easier in the light of a brief discussion of social causation. In contrast to neoclassical analysis we accept the principle of cumulative causation as the main hypothesis for the study of social relations and economic processes in particular.

The principle of cumulative or circular causation stresses the fact that social processes are marked by the interaction of several variables both "economic" and "noneconomic" which in their combined effects move the system away from a position of balance or equilibrium. In fact instead of calling forth a tendency toward automatic self-stabilization, social processes may be said to obey some principle of social inertia which tends to move the system in the same direction as the initial impulse. "The system is by itself not moving toward any sort of balance between forces but is constantly on the move away from such a situation.

In the normal case a change does not call forth countervailing change but, instead, supporting changes, which move the system in the same direction as the first change but much further." This principle of cumulative causation, by stressing the interaction of several factors that move the social system in the same direction as the initial impulse only faster constitutes the main causal hypothesis and the general conceptual framework for the study of social costs.

In a cumulative chain of causation there is necessarily interaction between many factors; this together with the fact that the final outcome of a cumulative social process always depends upon the concurrence of circumstances makes it possible to shift part of the causal "responsibility" to one or the other factor. While this will always be possible and while it may serve a purpose in a debate, it defeats the search for truth. The truth of the matter is that it is the whole cumulative process of unrestrained concentration of industries and the subsequent growth of urban communities which gives rise to the contamination of the atmosphere beyond the levels of concentration of pollutants that might be said to be compatible with human health. It is this process of unrestrained concentration, regardless of climate and topography, which gives rise to the social costs. Neither the private automobiles nor the chimneys of private dwellings in Donora nor, for that matter, the recurrent temperature inversion, nor the topography of the valley in which Donora is located, can reasonably be said to have given rise to air pollution and its harmful effects. What has given rise to air pollution is the unrestrained concentration of industries in this locality. What applies to Donora applies with equal force to Pittsburgh, to New York, London, Tokyo, the Meuse Valley, the cities of the Ruhr or the paper mills in Cornerbrook in Newfoundland, the mining towns of the Andes and may tomorrow apply to Calcutta or the emerging industrial centers in Africa. Indeed, it will apply to the industrial centers of the Soviet Union and China if concentration of industrial production is permitted to proceed without consideration of its harmful effects on human health and its social costs. That some of these social costs are shifted by one group of firms to other concerns (or for that matter cause additional outlays for the very firms which are contributing to the air pollution in the first place) in no way changes the social character of these costs. They are avoidable if organized society—that is, governments—will take the necessary steps and enforce rules which the situation demands.

Before turning to these practical measures it may be worthwhile to point out how the principles of business enterprise favor the emergence of the social costs of air pollution. The initial concentration of industrial production in a few centers, as indeed the location of industries in general under conditions of unlimited competition, will take place in accord ince with private cost-benefit calculations. Once established, the industry widens the market for a host of other industries; it offers employment and income opportunities to labor and capital; it provides a broader tax base for the emerging urban communities and the necessary public services. The locality becomes generally more attractive for additional investments, enterprise and labor and urban settlement. It is this expansionary momentum which serves to "polarize" industrial development in certain "nodal" centers, which soon gives rise to secondary and tertiary spread effects in the form of increasing outlets for agricultural products and consumers' industries in general. In the light of traditional economic theory, the process seems to proceed in harmony with the principle of social efficiency. For, after all, internal economies combine with external economies (in the narrow Marshallian sense) to make it appear rational to concentrate production in centers which are already established and offer some guarantee that the necessary social overhead investments (in roads, schools, communication) can be shared by a larger community. What is overlooked is that the concentration of industrial production may give rise to external diseconomies which may call for entirely new and disproportionate overhead outlays for which nobody may be prepared to pay. Thus by concentrating on the analysis of internal and external economies, and by stopping short of the introduction of the concept of social costs of unrestrained industrial concentration, traditional theory lends tacit support to the overall rationality of cumulative growth processes, no matter what their socially harmful effects may be. After all, what could be more "rational" than to exploit to the fullest extent the availability of internal and external economies. As long as social costs remain unrecognized and as long as we concentrate on costs that are internal to the firm or to the industry, we shall fail to arrive at socially relevant criteria.

It may be argued that, while the neglect of social costs may contribute to the cumulative growth process, it still would not explain the incomplete and inefficient process of combustion which gives rise to the emanation of pollutants into the atmosphere. For, obviously, if air pollution is a sign of inefficient and incomplete combustion of coal or oil, the question arises why would business enterprise permit such waste to continue? The answer is simply that what may be technologically wasteful might still be economical considering the fact that not only social costs can be shifted with impunity but, above all, that discounted private returns (or savings) obtainable from the prevention of the technological inefficiency and social costs may not be high enough to compensate for the private costs of the necessary abatement measures. The fact that the resulting pollution of the atmosphere may cause social costs far in excess of the costs of their abatement is not, and indeed cannot, be normally ex-

pected to be considered in the traditional cost-benefit calculations of pri
vate enterprise. According to one investigator, the costs of enforcing exist
ing air pollution ordinances (in Ontario) would involve expenditure o
only fifteen cents per capita, while the annual economic losses whic
could be saved in this way would amount to between ten and twenty
dollars per person.

THE SOCIAL COSTS OF WATER POLLUTION

As in the case of air pollution, those who are most directly responsible fo
the pollution of watercourses do not bear the adverse effects and mone
tary losses caused by their productive activities. As a result, they are
frequently not interested in minimizing these negative consequences. A
a matter of fact they tend to be interested rather in avoiding investment
for the necessary treatment of their waste products. By minimizing thei
internal costs they tend to shift and actually maximize the external o
social costs. Indeed, the costs of controlling water pollution either by
prior treatment of the materials discharged or by otherwise disposing o
them are likely to be much lower than the social costs arising from the
neglect of these measures. The damages caused by water pollution are
distributed over a great number of people who may be unable or unin
terested in bringing legal action against the main offender because it i
usually difficult and expensive to prove liability for damages in court
"Judicial precedent requires the demonstration of specific damage rathe
than general damage, and further requires quantitative estimates of the
amounts of damage experienced by specified individuals. Variations in
the natural quality of water and polluting substances make the proces
of marshalling such evidence lengthy and intricate." Furthermore, even i
liability and damages are proven it is also necessary to demonstrate tha
practicable means of abatement exist before an injunction may be ob
tained. This entails additional expense and further delay. Thus the dis
persion of damages and the excessive expenses for court action act as a
deterrent to the elimination of water pollution in modern industrial soci
eties. However, the main social costs of water pollution are not those sus
tained by individuals in the form of injuries to health or property o
recreational values, but rather those which arise from the depletion of a
major economic resource.

What would be a possible approach to the evaluation of social cost
of water pollution? The answer to this important question depends upor
the conceptual framework we choose. If we want to look upon wate
pollution and social costs only in terms of the benefit-cost calculus of busi

ess enterprise, we would have to try to arrive at a monetary value of he various losses and damages caused by the contamination of water-ourses. While this is possible in some instances, it would be a fruitless ndertaking in others. It is possible to calculate the market values of the ncreased costs of maintenance and repair of particular structures which re corroded prematurely by the chemicals discharged by an up-stream actory. Similarly, it would be possible to estimate the market value of he loss of livestock, the destruction of crops, the loss of soil fertility or ven of certain recreational facilities. Data of this sort are available. Iowever, they are far from satisfactory because they are not only in-omplete but misleading. They focus attention only on those values which an be estimated in dollars and cents. These data also tend to support the nisconception that the problem of water pollution is essentially a uniquely local blight" which can be corrected by general appeals to "the esponsibility" of business concerns and local taxpayers. Actually, as we ave endeavored to show throughout the preceding analysis, the problem f water pollution can be understood only if it is seen in the framework of ocial ecology and the threatening imbalance between a given supply f, and a rapidly increasing demand for, water in practically all parts of he world. Water and air pollution do much more than shift some of the osts of production to people living outside of a given area. They create new physical environment for man. Indeed, instead of the natural en-ironment in which man has lived for centuries, the permanent revolu-ion of technology has created a man-made environment the full implica-ions of which, for human health and human survival, are far from being ully understood. We are only at the threshold of the realization that this nan-made environment may be exceedingly detrimental to all life on this lanet. By far the greatest potential danger in this respect is to be found 1 the inadequate disposal of radioactive wastes and radioactive fallout. Vhat gives these hazards their unique and dramatic character are the umulative effects of exposure over the entire life span of the individual n his health and survival. It has been said that in public health, as in ther fields of human understanding, "we stand at the microchemical and nicrophysical frontier" beyond which there may lie the solution of many f the problems of the presently still incurable and chronic diseases. The umulative impact of a man-made environment on human health is the neasure of major and significant social costs. As already pointed out, we re just at the beginning of systematic inquiries which may ultimately rovide the basis for the elaboration of scientific standards and the eval-ation of the human costs of water pollution.

The threatening imbalance between an essentially constant supply nd a rapidly increasing demand for water opens up still another perspective of the social costs of water pollution. A dependable supply of clean water is essential to agricultural and industrial growth. Regions which are unable to conserve their water supply or which permit it to be polluted by human activities, destroy one of the productive forces upon which rest their present and future prosperity. The pollution of this resource, like the pollution of the atmosphere, is a social cost which needs to be fully assessed. Admittedly, these productive forces have social values which cannot be assessed in terms of either individual welfare criteria or market values. They are values-to-society for which the market calculus provides at best only a preliminary and certainly not an over-ruling yardstick. Social values of this sort call for an appraisal of overall social and political consequences of action or nonaction. In the specific case of water pollution, they call for knowledge of the cause and effects of the pollution of specific watercourses so that the consequences and comparative costs of different policies (including the policy of nonaction) can be stated and thus become the basis of rational political choice. Only systematic research conducted under impartial scientific auspices can provide the basis for estimating the social costs of water pollution with a reasonable degree of accuracy. Unless such a scientific assessment of the causes, extent and effects of water pollution is carried out, and kept up-to-date to take account of the effects of new pollutants resulting from a changing technology, it is futile to speak of overall measurements of the social costs of water pollution and of environmental pollutants in general. Such a scientific assessment would ultimately lead to the elaboration of scientific standards of Maximum Permissible Concentrations of various types of environmental contaminants. These standards could serve not only as the basis of measurements of environmental pollution but also as objective criteria for the formulation of sanitary standards and policies for the maintenance of an environment that conforms to the biological requirements of health, life and survival. Instead of passively adjusting to a detrimental environment, the scientific man can shape his surroundings and adjust his man-made (artificial) environment to his purposes.

We cannot concern ourselves with the economic or administrative aspects of abatement policies. Suffice it to point out, however, that the need for more information is not equivalent to saying that we must remain inactive until all investigations have been completed. Enough information is available to justify remedial action in specific cases. Delay will render not only future action more costly but will increase present social costs. Indeed, as pointed out before, the costs of early abatement may be amply justified if compared with the future social cost of neglecting early remedial action. While society can never avoid the social cost of neglect, individual enterprises can shift the social costs to others and

o the future. The social costs of water pollution are borne to a consider-
ble extent by people living downstream. This complicates the abatement
problem; it transforms water pollution into a regional problem. Above all,
t raises the question of the incidence of costs and benefits. Abatement
policies cannot be made dependent upon the local ability to pay or the
ocal tax base. They have to be financed on a regional scale and as part
of the general attempt to preserve a national resource. Only a regional
approach with federal agencies assuming overall responsibility for water
management policies can guarantee the preservation of water resources.

Effluents and affluence

JARED E. HAZELTON

This article describes the zoning approach used by the New England Interstate Water Pollution Control Compact. This regional authority classifies sections of rivers and streams according to a scale of potential water uses, balancing the cost of treatment and dumping restrictions against the benefits various kinds of water use bring to the community. The author is on the staff of the research department of the Federal Reserve Bank of Boston.

No other area of the country has been blessed with more abundan supplies of water than New England. Its rivers and streams originall provided the main setting for its economic development. They forme the principal travel routes, and naturally the towns and villages gre along the river banks. Rivers provided water-power to drive the earl mills and later supplied the large water requirements of the textil paper, tanning, brewing and other industries. At the same time, wate ways were a major recreational asset, providing opportunities for swin ming, boating and fishing as well as a habitat for wildlife.

Also from earliest times, the rivers and streams of New Englan were used for waste disposal. As long as water resources were plentifu relative to New England's needs, the region could afford the luxury c discharging untreated municipal and industrial wastes into its wate ways. Unfortunately, this is no longer the case. While industrial deman for water continue to grow only slowly, recreational demands have in creased rapidly and will undoubtedly soar in the future. More peopl now have more time for recreation and more money to spend. As a resu

Reprinted with permission from the *New England Business Review*, June 196 pp. 2–9.

inimum standards for water purity have risen. Pollution, therefore, is
more concern today than at any previous time.

Of course, not all pollution problems are sufficiently important to
stify treatment. Funds are limited and a balance must be achieved
between the costs to society and the benefits that will be derived.
hus the central task in combating pollution is to ensure that the avail-
le funds are spent where they will do the most good. This involves co-
rdinated planning between engineers and resource specialists as well as
between federal, state, and local authorities.

The New England states were among the first to develop a compre-
ensive plan for dealing with their water pollution problems. This article
ssesses the progress which has been made and the tasks remaining in
eaning up the region's waterways.

HE PROBLEM

istorically, pollution was regarded as a menace to the public health.
owever, as sanitation methods improved and public drinking supplies
egan to receive treatment, pollution ceased to be a major public health
roblem. Industries and municipalities continued to discharge untreated
astes into the waterways secure in the knowledge that their actions
id not threaten the public health.

As rising population increased the load of municipal wastes and as
conomic growth led to a rise in the number of industrial users of water,
new concern over pollution arose. New Englanders began to discover
at some public beaches had to be closed because of the high bacteria
unt of polluted waters. Anglers noticed that many stretches of water-
ay, formerly teeming with game fish, now were barren. In some in-
ances, pollution became so severe as to offend the senses, causing pun-
ent odors and unsightly scum and virtually killing all fish and wildlife
bout the stream.

Thus pollution came to be recognized as preventing many uses of
ater. It is this concept of pollution which provides the key to an eco-
omic definition of the problem. Pollution must be related not only to
e type and amount of wastes being discharged into the waterway, but
lso to the effect of such discharges in unreasonably impairing the water's
alue for a variety of uses.

ETTING STANDARDS

ollution stems primarily from the discharge of waste from two sources:
ewage from municipalities and various types of waste from industry.

Since the total amount and type of waste going into a river and the tot capacity of the river to absorb waste determine the pollution, its contr must be planned on a broader basis than the individual municipality industry. The planning unit is therefore the state on intrastate strear and a collection of states on interstate waterways.

Clearly the control of pollution must be related to the functions t the water is to serve. Minimum treatment ensures that water has objectionable odor or appearance. The extent of treatment must be pr gressively increased if water is to be suitable for boating, fishing, bathi and, finally, drinking.

One method of control would be to require municipalities and i dustries to treat their water until it is fit for the highest possible use that is, drinking. The costs of such a program, however, would run in billions of dollars. Furthermore, where the water is not needed as a drin ing supply, the costs would certainly exceed the benefits of treatment.

A more realistic approach to pollution abatement is to classi waterways according to the highest use to which particular segmer should be put, balancing costs against benefits. Costs depend on t capacity of the river to assimilate wastes, the nature of the wastes bei discharged, and the desired degree of treatment. Benefits, on the oth hand, require judgment as to the value of increased treatment over t whole range of water uses. Once made, classifications can be either for ally adopted by the state or interstate group, or they may be used guides by control agencies in carrying out their enforcement role.

It is obvious that some type of classification or setting of standar is essential in any water pollution control program, since enforcement a tion must be related to standards of use. However, the danger in esta lishing formal classifications is that once made they are difficult to chang and that waterways already heavily polluted tend to be classified as me carriers of waste.

Moreover, judgments as to the reasonable uses of water may diffe For example, a community faced with choosing between constructing new school or a waste treatment plant might not view its pollution in t same light as a downstream user. Classification helps resolve these co flicts of interest by providing a comprehensive plan for the whole river.

STATUS OF POLLUTION CONTROL
IN NEW ENGLAND

Classification

The adoption of formal classifications for waterways has been use more extensively in New England than in any other area of the natio

he majority of interstate rivers in the region have already been classi-
ed. Studies remain to be completed only on portions of the Androscog-
in River (New Hampshire–Maine), the Connecticut River (New Hamp-
iire–Vermont), Lake Champlain Basin (Vermont), and the Merrimack
iver (New Hampshire).

Classification has been carried out under the auspices of the New
ngland Interstate Water Pollution Control Compact. This compact,
ormed in 1947, includes the eastern portion of New York as well as the
x New England states, and is administered by the New England Inter-
ate Water Pollution Control Commission. Members of the Commission
re from each of the states, are appointed by their governors, and include
ate officials concerned with health and water pollution as well as other
tizens representing municipal, industrial, and recreational interests. The
ommission passes on the classifications proposed by the affected states.
fter classification has been approved, each state holds the responsibility
r obtaining action by municipalities and industries for the installation
f waste treatment works to meet the classification requirements.

The classifications used by the New England States range from
lass A, water suitable for any use, to Class D, water suitable for trans-
ortation of sewage and industrial wastes without nuisance, and for
ower, navigation, and certain industrial uses. Water failing to meet even
iese last requirements is termed Class E, unsatisfactory or objectionable.

Construction of Treatment Plants

One gauge of progress in water pollution control is the number of
iunicipal and industrial treatment plants constructed. In the last seven-
een years, the proportion of the sewered population of New England
erved by sewage treatment plants has grown from 39 to 67 percent. With
ie completion of treatment plants now under construction or about to be
tarted, 87 percent of the sewered population will be served. Moreover,
ommunities comprising an additional 11.5 percent of the sewered popu-
ition have had preliminary engineering plans made for sewage treat-
ient plants. This leaves only 1.5 percent of the sewered population who
ave done no planning for sewage treatment facilities.

Since 1947, $330 million have been spent on sewage works in New
ngland. During 1964, works costing over $15 million were completed
nd included eight new treatment plants. Construction was continued on
rojects estimated to cost $38 million and was begun on works totaling
29 million.

About three-fourths of the 1900 waste-producing industrial plants
1 the New England compact area are estimated to have private treat-
ient facilities or to discharge to municipal sewer systems which now

have or eventually will have treatment works. In the past seventee years, over 450 industrial plants have installed waste treatment faciliti and approximately the same number have connected to municipal sewe systems.

Present Stream Conditions

The number of municipal and industrial treatment plants co structed is, however, only a limited measure of progress in pollution co trol. There are many degrees of treatment. Primary treatment, the lev most often provided, removes only about 50 percent of the pollutants fro the discharge. Further, in many instances, the remaining industri sources of pollution represent "hard core" problems where treatment either very difficult or extremely expensive. Finally, some growth in co struction of treatment plants is required merely to accommodate increas in population and the extension of municipal sewer lines. Thus thes plants do not improve the present status of pollution.

A broader gauge of progress would be the number of polluted rive restored to acceptable quality. In these terms, present conditions are f below even the modest goals set for the region under the classificatior which have been adopted. Much remains to be done before a significar portion of New England waters will be restored even to moderate qualit standards.

PROSPECTS FOR THE FUTURE

Increased Federal Role

Until recently, the U.S. Congress has regarded pollution abatemer largely as the province of the states. In 1956, however, Congress passe the Federal Water Pollution Control Act. This act, while recognizing tha the primary responsibility for pollution control rests with the states, pre vided for federal assistance to communities for construction of municipe sewage treatment works. Annual appropriations of $90 million for th fiscal year 1963 and $100 million for each of the next four years were au thorized. The grants cannot exceed 30 percent of the project cost c $600,000, whichever is the smaller. Grants to the New England State under this law in fiscal 1964 totaled more than $7 million.

In 1961, the act was amended to strengthen the federal goverr ment's power to intervene in interstate or navigable waters affecting ir terstate conditions. The amendments empowered the federal governmer

*Classification and Standards of Quality for Interstate
Waters (As Revised and Adopted October 1, 1959)*

lass A	Suitable for any water use. Character uniformly excellent.
ass B	Suitable for bathing and recreation, irrigation and agricultural uses; good fish habitat; good aesthetic value. Acceptable for public water supply with filtration and disinfection.
ass C	Suitable for recreational boating, irrigation of crops not used for consumption without cooking; habitat for wildlife and common food and game fishes indigenous to the region; industrial cooling and most industrial process uses.
ass D	Suitable for transportation of sewage and industrial wastes without nuisance, and for power, navigation, and certain industrial uses.

initiate enforcement actions in any of three situations: 1. when a state quests such action on interstate pollution; 2. when the federal government believes that interstate pollution is endangering the health or welre of another state; and 3. when a governor believes that interstate pollution is a danger to persons within his state.

Formal actions under this law consist of conferences, public heargs, warnings to specific polluters to take remedial steps determined by e federal government, and court action if necessary. Over thirty such ctions have been undertaken thus far, with only one reaching the court age. Four conferences have been held in New England: on the Androoggin and the Connecticut in 1963; the Merrimack in 1954; and the lackstone-Ten Mile in 1965. None of these New England conferences sulted in any further federal action. In general, the conferences conuded that the state programs were adequate and urged accelerated forts to carry them out.

The federal government will probably play an increasingly important role in pollution abatement efforts in the future. In addition to the rong commitment of the present administration, both houses of Conress this year have passed an amendment to the Federal Water Pollution Control Act. While a Joint Congressional Conference Committee is resently ironing out the differences in these bills, indications are that an ct will pass the Congress during this session.

The new act is expected to increase the amount of federal funds vailable to communities for the construction of waste treatment plants. would also provide for the establishment of a new agency under the ecretary of Health, Education and Welfare which will have responsibily for all federal pollution abatement programs. In all probability, fedral enforcement powers will be strengthened, perhaps to include the

right to establish classification for rivers where states have not done ,
and to review existing classifications.

State Responsibilities

With the federal government expressing a greater interest in wat
pollution control, it is likely that the states will come under increasi
pressure to step up their pollution abatement programs.

The New England states have a start since work on the classificati
of the interstate waterways has been carried on during the past sevente
years. This aspect of the work is expected to be completed in the ne
few years. The remaining major task will be to accelerate construction
waste treatment plants to bring the waterways up to classification leve
Pressure will therefore be transferred to localities and industries sin
they hold the primary responsibility for action. While a concerted effc
to build treatment plants has accompanied classification efforts, parti
ularly since 1957, many more plants will be required in the future if tl
classifications shown on the map are to be met.

To encourage the completion of pollution abatement facilities, tl
states combine the carrot of grants-in-aid for communities building trea
ment plants with the stick of court enforcement for the communities ar
industries that lag behind. Vermont now provides municipalities wi
grants-in-aid of 20 percent of the cost of construction, while Maine h
recently increased its contribution to 30 percent. New Hampshire gran
its communities 30 percent of the annual cost of financing treatme:
plants. It seems likely that more state funds will be made available f
these purposes in the next few years. For example, the Massachuset
General Court is now considering an air proposal similar to that in Ne
Hampshire.

Controls Needed

Historically, the danger of flooding and the existence of water pc
lution have prevented development of much of the land abutting tl
major rivers and streams of New England. As a result, this region has a
inheritance of scenically attractive waterways moving through some of i
most densely populated areas.

All this may soon change. Over the next ten years, the federal, sta
and local governments in New England will spend more than $500 m
lion on municipal sewage treatment facilities. During this period abo
$100 million will also be spent on flood control. The resulting enhanc

ent of values on the land abutting our major rivers will likely encour-
ge its development. This may be a mixed blessing, however.

While pollution destroys many of the potential recreational uses for
ater, abatement of pollution does not in itself make possible these uses.
o the contrary, by enhancing the value of waterfront property, pollution
ntrol may encourage a type of development which is just as incom-
atible with bathing, fishing, and boating and just as destructive of scenic
lues as pollution itself.

Imaginative planning and foresight are required if abatement of
ater pollution is not to lead to the growth of other forms of environ-
ental pollution. Speeding motorboats, for example, can prevent or limit
athing, canoeing, sailing, and other recreational opportunities. Riverside
onky-tonks, garish neon signs, and other types of thoughtless develop-
ent can destroy the scenic value of an otherwise attractive river. Lack
 public access can restrict the usefulness of waterways in fulfilling de-
ands for outdoor recreation.

Many of these problems would be solved if state and local govern-
ents assumed greater responsibility for influencing the type of private
evelopment by zoning the land and restricting the use of the waterways.
t the same time, they could also purchase fishing and access rights to
reams and provide for new parks, playgrounds, and wildlife refuges
ong the riverways.

Only in this way could public investment in pollution abatement
cilities be made to yield broad-scale public benefits. The added costs
 these measures would be small compared to the large investment re-
ired for pollution abatement facilities. The potential benefits would be
eat, particularly in light of the ever-growing demand for outdoor rec-
ation and the limited open areas available to the public.

Economic incentives i
air-pollution contro

EDWIN S. MILLS

Edwin S. Mills, a professor of economics at Johns Hopkins University, has been on the faculty of the Massachusetts Institute of Technology, and recently served as senior staff economist of the Council of Economic Advisers. He is a specialist in price theory. In this essay he shows how economic theory can be adapted to cope with air pollution.

Smoke is one of the classic examples of external diseconomies me
tioned in the writings of Alfred Marshall and his followers. Generatio
of college instructors have used this form of air pollution as an illustr
tion to help their students to understand conditions under which compe
tive markets will or will not allocate resources efficiently. By now, t
theoretical problems have been explored with the sharpest tools availal
to economists. The consensus among economists on the basic issue is ov
whelming, and I suspect one would be hard-pressed to find a propositi
that commands more widespread agreement among economists than t
following: The discharge of pollutants into the atmosphere imposes
some members of society costs which are inadequately imputed to t
sources of the pollution by free markets, resulting in more pollution th
would be desirable from the point of view of society as a whole.

In spite of the widespread agreement on the fundamental issu
regarding externalities such as air pollution, there have been remarkal
few attempts in the scholarly literature to carry the analysis beyond t
point. Most writers have been content to point out that the free mar
will misallocate resources in this respect, and to conclude that this jus

Reprinted with permission from *The Economics of Air Pollution*, edited
Harold Wolozin, © 1966 by W. W. Norton & Company, New York.

es intervention. But what sort of intervention? There are many kinds, nd some are clearly preferable to others.

Too often we use the imperfect working of a free market to justify ny kind of intervention. This is really an anomalous situation. After all, narkets are man-made institutions, and they can be designed in many ays. When an economist concludes that a free market is working badly giving the wrong signals, so to speak—he should also ask how the mar- t may be restructured so that it will give the right signals.

Thus, in the case of air pollution, acceptance of the proposition ated above leads most people to think entirely in terms of direct regula- on—permits, registration, licenses, enforcement of standards and so on. submit that this is rather like abandoning a car because it has a flat tire. f course, in some cases the car may be working so badly that the pres- nce of a flat tire makes it rational to abandon it, and correspondingly e inadequacies of some market mechanisms may make abandonment esirable. Nevertheless, I submit that the more logical procedure is to ask ow a badly functioning market may be restructured to preserve the clear dvantages of free and decentralized decision-making, but to remedy its efects. Only when there appears to be no feasible way of structuring a narket so that it will give participants the right signals, should it be ven up in favor of direct regulation.

It is easy to state the principle by which the socially desirable nount of pollution abatement should be determined: *Any given pollu- on level should be reached by the least costly combination of means vailable; the level of pollution should be achieved at which the cost of a irther reduction would exceed the benefits.*

To clothe the bare bones of this principle with the flesh of substance a very tall order indeed. In principle, if every relevant number were nown, an edict could be issued to each polluter specifying the amount y which he was to reduce his discharge of pollutants and the means by hich he was to do so. In fact, we are even farther from having the right umbers for air pollution than we are from having those for water pollu- on.

In this situation, I suggest that any scheme for abatement should be onsistent with the following principles:

1. It should permit decision-making to be as decentralized as pos- ble. Other things being equal, a rule that discharges must be reduced y a certain amount is preferable to a rule that particular devices be in- alled, since the former permits alternatives to be considered that may e cheaper than the devices specified in the latter.

2. It should be experimental and flexible. As experience with abate-

ment schemes accumulates, we will gain information about benefits an
costs of abatement. We will then revise our ideas about the desirab
amount and methods of abatement. Control schemes will have to b
revised accordingly.

3. It should be coupled with careful economic research on benefi
and costs of air-pollution abatement. Without benefit-cost calculation
we cannot determine the desirable amount of abatement. We can, how
ever, conjecture with confidence that more abatement is desirable tha
is provided by existing controls. Therefore, our present ignorance of be
efits and costs should not be used as an excuse for doing nothing. I wou
place great emphasis on doing the appropriate research as part of an
control scheme. A well-designed scheme will provide information (e.g
on the costs of a variety of control devices) that is relevant to the be
efit-cost calculations.

MEANS OF CONTROL

We are not in a position to evaluate a variety of schemes that are in u
or have been proposed to control or abate air pollution. It will be usef
to classify methods of control according to the categories employed l
Kneese in his discussion of water pollution:

1. *Direct Regulation.* In this category, I include licenses, permit
compulsory standards, zoning, registration, and equity litigation.

2. *Payments.* In this category I include not only direct payments a
subsidies, but also reductions in collections that would otherwise b
made. Examples are subsidization of particular control devices, forgiv
ness of local property taxes on pollution-control equipment, accelerate
depreciation on control equipment, payments for decreases in the di
charge of pollutants, and tax credits for investment in control equipmen

3. *Charges.* This category includes schedules of charges or fees fo
the discharge of different amounts of specified pollutants and excise a
other taxes on specific sources of pollution (such as coal).

My objection to direct regulation should be clear by now. It is to
rigid and inflexible, and loses the advantages of decentralized decisio
making. For example, a rule that factories limit their discharges of po
lutants to certain levels would be less desirable than a system of efflue
fees that achieved the same overall reduction in pollution, in that th
latter would permit each firm to make the adjustment to the extent an
in the manner that best suited its own situation. Direct restrictions ar

ually cumbersome to administer, and rarely achieve more than the
'ossest form of control. In spite of the fact that almost all of our present
)ntrol programs fall into this category, they should be tried only after
I others have been found unworkable.

Thus, first consideration ought to be given to control schemes under
ie second and third categories.

Many of the specific schemes under these two categories are un-
:sirable in that they involve charges or payments for the wrong thing.
it is desired to reduce air pollution, then the charge or payment should
:pend on the amount of pollutants discharged and not on an activity
at is directly or indirectly related to the discharge of pollutants. For
:ample, an excise tax on coal is less desirable than a tax on the discharge
' pollutants resulting from burning coal because the former distorts re-
urce use in favor of other fuels and against devices to remove pollu-
nts from stack gases after burning coal. As a second example, a pay-
ent to firms for decreasing the discharge of pollutants is better than a
x credit for investment in pollution-control devices because the latter
troduces a bias against other means of reducing the discharge of pollu-
nts, such as the burning of nonpolluting fuels. Thus, many control
:hemes can be eliminated on the principle that more efficient control can
)rmally be obtained by incentives that depend on the variable it is
:sired to influence rather than by incentives that depend on a related
iriable.

Many of the specific schemes under *Payments* can be eliminated on
ie grounds that they propose to subsidize the purchase of devices that
:ither add to revenues nor reduce costs. Thus, if a pollution-control de-
ce neither helps to produce salable products nor reduces production
)sts, a firm really receives very little incentive to buy the device even if
ie government offers to pay half the cost. All that such subsidy schemes
:complish is to reduce somewhat the resistance to direct controls. Of
)urse, some control devices may help to recover wastes that can be made
ito salable products. Although there are isolated examples of the recov-
y of valuable wastes in the process of air-pollution control, it is hard to
1ow whether such possibilities are extensive. A careful survey of this
ibject would be interesting. However, the key point is that, to the ex-
:nt that waste recovery is desirable, firms receive the appropriate in-
:ntive to recover wastes by the use of fees or payments that are related
> the discharge of effluents. Therefore, even the possibility of waste re-
)very does not justify subsidization of devices to recover wastes.

The foregoing analysis creates a presumption in favor of schemes
nder which either payments are made for reducing the discharge of pol-
itants or charges are made for the amount of pollutants discharged. The

basic condition for optimum resource allocation can in principle be satisfied by either scheme, since under either scheme just enough incentive can be provided so that the marginal cost of further abatement approximates the marginal benefits of further abatement. There are, however, three reasons for believing that charges are preferable to subsidies:

1. There is no natural "origin" for payments. In principle, the payment should be for a reduction in the discharge of pollutants below what it would have been without the payment. Estimation of this magnitude would be difficult and the recipient of the subsidy would have an obvious incentive to exaggerate the amount of pollutants he would have discharged without the subsidy. The establishment of a new factory would raise a particularly difficult problem. The trouble is precisely that which agricultural policy meets when it tries to pay farmers to reduce their crops. Jokes about farmers deciding to double the amount of corn not produced this year capture the essence of the problem.

2. Payments violate feelings of equity which many people have on this subject. People feel that if polluting the air is a cost of producing certain products, then the consumers who benefit ought to pay this cost just as they ought to pay the costs of labor and other inputs needed in production.

3. If the tax system is used to make the payments, e.g., by permitting a credit against tax liability for reduced discharge of pollutants, a "gimmick" is introduced into the tax system which, other things being equal, it is better to avoid. Whether or not the tax system is used to make the payments, the money must be raised at least partly by higher taxes than otherwise for some taxpayers. Since most of our taxes are not neutral, resource misallocation may result.

I feel that the above analysis creates at least a strong presumption for the use of discharge or effluent fees as a means of air-pollution abatement.

Briefly, the proposal is that air pollution control authorities be created with responsibility to evaluate a variety of abatement schemes, to estimate benefits and costs, to render technical assistance, to levy charges for the discharge of effluents, and to adopt other means of abatement.

Serious problems of air pollution are found mostly in urban areas of substantial size. Within an urban area, air pollution is no respecter of political boundaries, and an authority's jurisdiction should be defined by the boundaries of a metropolitan air shed. Although difficult to identify precisely, such air sheds would roughly coincide with Standard Metropolitan Statistical Areas. Except in a few cases, such as the Chicago Gary

d the New York–northern New Jersey areas, jurisdiction could be con-
ed to a single metropolitan area. In a number of instances, the author-
y would have to be interstate. In many large metropolitan areas, the
thority would have to be the joint creation of several local govern-
ents. There would presumably be participation by state governments
d by the federal government at least to the extent of encouragement
d financial support.

Each authority would have broad responsibility for dealing with air
ollution in its metropolitan air shed. It would institute discharge fees
d would be mainly financed by such fees. It would have the respon-
bility of estimating benefits and costs of air-pollution abatement, and
setting fees accordingly. It would have to identify major pollutants in
s area and set fees appropriate to each significant pollutant. The author-
y could also provide technical advice and help concerning methods of
atement.

Although there would be great uncertainty as to the appropriate
vel of fees at first, this should not prevent their use. They should be set
nservatively while study was in progress, and data on the responses of
ms to modest fees would be valuable in making benefit-cost calcula-
ons. Given present uncertainties, a certain amount of flexible experi-
entation with fees would be desirable.

Questions will necessarily arise as to just what kinds and sources of
ollutants would come under the jurisdiction of the proposed authority. I
o not pretend to have answers to all such questions. Presumably, stand-
d charges could be set for all major pollutants, with provision for varia-
on in each metropolitan air shed to meet local conditions. It is clear
at provision should be made for the possibility of varying the charge
r a particular pollutant from air shed to air shed. The harm done by the
ischarge of a ton of sulfur dioxide will vary from place to place, de-
ending on meteorological and other factors. It is probably less harmful
Omaha than in Los Angeles. It is important that charges reflect these
ifferences, so that locational decisions will be appropriately affected.

Consideration would also have to be given to the appropriate tem-
oral pattern of charges. In most cities, pollution is much more serious in
mmer than at other times. Charges that were in effect only during sum-
er months might induce a quite different set of adjustments than
arges that were in effect at all times.

No one should pretend that the administration of an effective air-
ollution control scheme will be simple or cheap. Measurement and
onitoring of discharges are necessary under any control scheme and can
e expensive and technically difficult. Likewise, whatever the control
heme, finding the optimum degree of abatement requires the calcula-

tion of benefits and costs; these calculations are conceptually difficult an demanding.

The point that needs to be emphasized strongly is that the cost administering a control scheme based on effluent fees will be less tha the cost of administering any other scheme of equal effectiveness. A effluent-fee system, like ordinary price systems, is largely self-administe ing.

This point is important and is worth stating in detail. First, con sider an effluent-fee system. Suppose a schedule of fees has been se Then firms will gradually learn the rate of effluent discharge that is mo profitable. Meanwhile, the enforcement agency will need to sample th firm's effluent to ensure that the firm is paying the fee for the amount a tually discharged. However, once the firm has found the most profitab rate of effluent discharge, and this is known to the enforcing agency, th firm will have no incentive to discharge any amount of effluent other tha the one for which it is paying. At this point the system becomes self-a ministering and the enforcement agency need only collect bills. Secon consider a regulatory scheme under which the permissible discharge set at the level that actually resulted under the effluent-fee scheme. The the firm has a continuing incentive because of its advantage on the cos side to exceed the permissible discharge rate so as to increase productio Monitoring by the enforcement agency therefore continues to be ne essary.

Of course, under either a regulatory or an effluent-fee scheme, change in conditions will require the search for a new "equilibrium Neither system can be self-enforcing until the new equilibrium has bee found. The point is that the effluent-fee system becomes self-enforcing that point, whereas the regulatory system does not.

The economics of
environmental quality

SANFORD ROSE

The ultimate solution to environmental disruption is recycling. All too often, most people fail to realize that there is no such thing as throwing away something. Everything reenters the ecological system in some form or other. Sanford Rose, an associate editor of Fortune *examines in this article how economic incentives can be designed to stimulate more constructive forms of recycling.*

Arguments over pollution control often exhibit the fallacy of all or othing. Some people seem to feel that the environment should be reored to pre-industrial purity. Others apparently agree with the mayor of smallish midwestern city who recently told a citizens' group: "If you ant the town to grow, it's got to stink." Neither viewpoint is acceptle.

Those who imagine that pollution can be totally eliminated fail to rasp the dimensions of the waste problem. As some economists have reently suggested, it might be well to dispose of the expression "final conumption." People and businesses do not consume goods; they extract tilities from goods before discarding them. Such things as gems, works f art, heirlooms, and monuments might be thought of as being, in some nse, finally consumed. But all other goods—durables, nondurables, and yproducts—are eventually either discharged to the environment or cyed back into the production process. About 10 to 15 percent of total utput, however, is temporarily accumulated in the form of personal posssions, capital goods, additions to inventory, etc. If society were to op accumulating for a while, observes economist Allen V. Kneese of

149

Resources for the Future, the weight of residuals discharged into th natural environment would equal the weight of raw materials used plu the weight of the oxygen absorbed during production. In other word waste disposal would be an even larger operation than basic materi production.

Though waste and pollution problems are too big to be eliminate they can be ameliorated by producing fewer goods (or a different mix c goods), by recycling more of what has been produced, or by changing th form of wastes or the manner of their disposal. These alternatives ar subject to economic evaluation. In principle, pollution is at an optimi level when the cost of additional amelioration would exceed the benefit If by spending a dollar an upstream mill can save downstream wate users at least a dollar, it should do so—from society's point of view.

TOO MUCH PAPER, TOO LITTLE FISH

Unfortunately, until very recently upstream mills were profoundly disir clined to spend anything to relieve downstream distress. They were ac customed to regarding the waste-disposal capacity of the stream as free good. But wastes discharged upstream can impose costs of one kin or another downstream. Such costs are labeled, among other thing spillovers, side effects, external diseconomies, disamenities, and exter nalities.

One important effect of externalities is to warp the allocation c productive resources. Because a paper mill, for example, can get by with out cleaning up its wastes, its costs of production are artificially under stated. Since in a competitive economy prices tend to reflect productio costs, the mill's prices may also be understated. If so, the result is greate demand for paper than if prices reflected the *full costs* of paper produc tion—both the costs borne internally by the mill and those borne exter nally by the mill and those borne externally downstream. At the sam time, some downstream producers—fisheries, perhaps—may have highe costs and prices, tending to depress demand. Society thus gets relativel too much paper and too little fish, and consumers of fish in effect sul sidize consumers of paper. In some degree, then, resources are allocate with less than maximum efficiency and equity.

Externalities can be reduced (never eliminated) by many differer means, including environmental standards, taxes, charges, subsidies, an generalized pressure. Each strategy results in a different mix of industria municipal, and federal expenditures on environmental quality.

ᵦIASES UP, BIASES DOWN

᠄ike any other investments, outlays for environmental improvement can
ᵦe evaluated by standard benefit-cost analysis. Where capital outlays are
ᵣequired, as would usually be the case, the basic budgeting procedure is
ᵣo forecast, for each year of the project's life, the probable benefits (dam-
᠎ges avoided) minus the operating costs. Since it is a truism that a dollar
᠎arned next year is worth less than a dollar in hand, these net benefit
᠎evels must be discounted to their present value. Normally this is done
ᵦy multiplying each year's net benefits by a discount factor based upon
ᵗhe estimated "opportunity cost" of capital—that is, what could have been
᠎arned if the funds had been used differently.

If the sum of the discounted net benefits exceeds the present cost of
ᵃacilities, the project is economically sensible. If discounted benefits fall
ᵗhort of present costs, the project should be rejected (according to ra-
ᵗional economics) because a dollar spent on further control would yield
᠎ess than a dollar's worth of damage abatement—i.e., the marginal costs
ᵥould exceed the marginal benefits.

In assessing the benefits and costs of pollution-control projects, gov-
᠎rnment inevitably finds itself in a statistical scissors. Those that will
ᵣave to pay for environmental improvements, such as businesses and
ᵐunicipalities, tend to inflate costs and deflate benefits. Those who par-
ᵗicularly want the improvements—recreationists, let us say—can be
᠎ounted on to do the reverse.

Many economists are convinced that of the two biases, the recrea-
ᵗionists' happens to be the right one. Respectable project analyses, they
ᵣrgue, are typically biased in favor of rejection, because costs are over-
ᵗated and benefit levels understated. Overstatement of costs is traceable
ᵣo the human tendency to travel familiar roads. When project analysts
ᵗalk about abating water pollution, for example, they usually mean con-
ᵗructing plants for secondary or tertiary treatment of effluents. But in
᠎ome parts of the country it would be much cheaper not to treat waste
ᵥater at all, but simply pipe it to storage lagoons for settling. Eventually
ᵗhe waste water could be used for irrigation. In other areas, costs can be
᠎reatly reduced by supplementing waste treatment with modification of
ᵖroductive inputs, changes in production processes, artificiaᵗ aeration of
ᵗtreams, augmentation of low stream flow by planned releases from res-
᠎rvoirs, and storage of wastes for eventual discharge during periods of
ᵣigh flow. When the project analyst fails to scan the full range of tech-

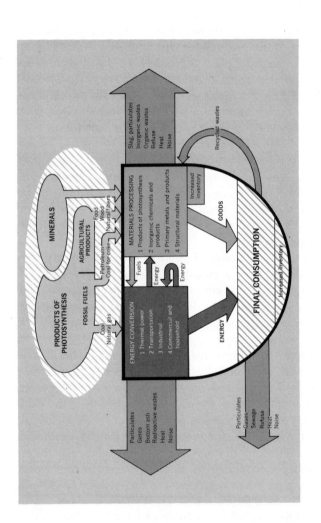

You can't get rid of matter, according to a well-known law of physics. All you can do is transform it. Modern economies, like that of the U.S., are good at taking the concentrated and transforming it into the diffuse; they are not so good at doing the opposite. It is easy to turn coal into pollutants such as fly ash, gases, and soot, but difficult—economically, if not technologically—to turn the fly ash back into, say, cinderblocks. But we have to find ways to slim down those thick pollution arrows and fatten up that skinny recycled-wastes arrow. This diagram of material flows in the economy is adapted from a concept worked out by economist Allen V. Kneese and physicist Robert U. Ayres. Intermediate goods that are neither discarded nor used go into material-processing inventory, distinct from final-consumption inven-

ological options—and he usually does fail—he is bound to come up with
n outsize price tag.

The benefits of environmental improvement, on the other hand, tend
o be understated because whole categories of damages are generally
ossed out of the calculation. The researcher usually concentrates on
measuring physical damages. (This is hard enough because the relationships between quantity of pollutants and resultant damage are both complex and highly variable.) If he is at all sensitive to the skepticism of his
colleagues, the researcher will make little or no attempt to quantify non-
physical damages—such as the impairment of human effectiveness or well-
being resulting from air pollution, or the loss of recreation and aesthetic
values resulting from water pollution.

Some respected pollution economists, indeed, have abandoned sys-
ematic damage estimation altogether. They point out that governmental
tandards for air and water quality are in the process of being estab-
ished. Since damages cannot be measured with either precision or com-
prehensiveness, these standards are in a sense arbitrary and suboptimal—
hat is, they may overcontrol or undercontrol pollution. But for better or
worse, the argument runs, the standards are there, and economists should
devote their efforts to working out least-cost ways of satisfying the new
criteria.

As a result, most estimates of pollution damage are elaborate
guesses. For example, the most frequently quoted figure for total annual
air-pollution damage in the U.S.—$11 billion—is really an estimate of
cleaning costs derived from smoke-damage data for Pittsburgh in 1913. In
hat year, investigators for the Mellon Institute calculated that Pittsburgh's smoke nuisance imposed additional costs—in cleaning and laun-
dering clothes and maintaining and lighting homes, businesses, and public
buildings—amounting to $20 a year per person. Many years later, this $20
figure was adjusted to 1959 prices on the basis of the commodity price
ndex. The updated per capita damage estimate was then multiplied by
he 1958 U.S. population to arrive at $11 billion.

One major omission from this obviously shaky figure is the cost of
air-pollution effects on human health. In an as yet unpublished paper,
economists Lester B. Lave and Eugene Seskin of Carnegie-Mellon University have made a serious statistical effort to assess this cost. Working
from a variety of medical sources, Lave and Seskin correlated differences
n mortality and disease rates in several geographic areas with differences
n social class, population density, and two indices of air pollution. It
urned out that air pollution and the bronchitis death rate were signif-
cantly correlated. According to these findings, reducing the amount of
pollution in all geographic areas to the level of the cleanest region stud-

ied would lower the bronchitis death rate by between 40 and 70 percent
Other analyses revealed significant associations between air pollution and
heart disease, emphysema, lung cancer, and infant mortality.

Many of the medical studies that Lave and Seskin used have been
criticized on the grounds that the level of pollution may be correlated
with unmeasured factors that are the real causes of ill health. For example, some of the studies do not control for differences in occupational exposure to pollution, in smoking habits, and in the general pace of life
Lave and Seskin took pains to try to meet this kind of objection. "It i
especially hard," says their research paper, "to believe that air pollution
and ill health are spuriously related when significant effects are found
comparing individuals within strictly defined occupational groups, such
as postmen or bus drivers (where incomes and working conditions are
comparable and unmeasured habits ought to be similar)."

Lave and Seskin argue that roughly 25 percent of all respiratory disease is associated with air pollution. Therefore pollution must account for
one-fourth of all the costs, both direct (mainly hospital and doctor bills
and indirect (basically forgone earnings). Estimating conservatively, they
say, the two economists find that the health damages attributable to air
pollution amounted to $2 billion in 1963, the last year for which usable
cost data are available. This is one of the biggest estimates to date
Ronald Ridker, a pollution economist now working for AID, put the 1958
costs of all respiratory disease at $2 billion (again, on a conservative reck-
oning) and the portion attributable to air pollution at about $400 million

DISCOUNTS FOR AILING SHRUBBERY

To carry out an adequate economic analysis of air-pollution abatement
—or any undertaking to improve environmental quality—the analyst has
to relate benefits to costs. More precisely, he has to estimate the specific dollar value of benefits to be derived from a given additional expenditure. A study done by Professor Thomas D. Crocker of the University of Wisconsin and Professor Robert J. Anderson Jr. of Purdue make
it possible to perform a marginal analysis of this kind in the air-pollution
field. They argue that some of the damages associated with air pollution
are reflected in property values. The potential home buyer perceives and
evaluates many of the effects of air pollution on residential property—e.g.
ailing shrubbery, off-color paint, sooty surfaces, unpleasant odors, hazy
view, etc. And whether or not he connects such disamenities with air
pollution, the buyer will take them into account in his offer price.

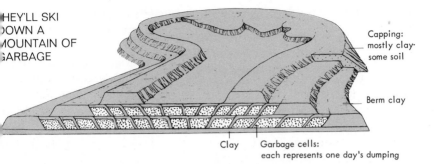

THEY'LL SKI DOWN A MOUNTAIN OF GARBAGE

Capping: mostly clay; some soil

Berm clay

Clay   Garbage cells: each represents one day's dumping

Mount Trashmore, as it is frequently called in Du Page County, Illinois, is a hill slowly being sculpted out of garbage. Located in the Blackwell Forest Preserve, just west of Chicago, the hill will eventually reach 125 feet, the highest elevation in the county. It will feature six toboggan runs and five ski slopes in an area short on facilities for winter recreation. Mount Trashmore was conceived by John Sheaffer, a waste-management specialist at the University of Chicago. Du Page County officials came to him with two problems: they didn't know what to do with a badly scarred marshy pit, known as the Badlands, and they were running out of available landfill space for garbage. Sheaffer proposed excavation below the water table to turn the pit into a lake and use of the excavated clay to help build a mountain of garbage. The clay would form an impermeable barrier upon which garbage could be stacked without danger of groundwater contamination. The project, which started in 1965, is cheap: the sale of gravel excavated from the pit will cover much of the cost. The mountain will be a honeycomb of cells, each four feet deep—three feet of garbage, one foot of clay. Each layer of cells is surrounded by a thick clay wall, or berm. County officials call their mountain, to be finished late this year, "the seventh engineering wonder of the world."

Anderson and Crocker tested their hypothesis statistically through an analysis of residential property values in three cities—St. Louis, Washington, and Kansas City. They correlated variations in sale prices and rents with family income, number of rooms, age and condition of property, distance from the center of the city, racial composition of neighborhood, and general educational level of neighborhood, as well as with two pollution variables—sulphur trioxide and suspended particulates. Air pollution and property values proved to be inversely related to a significant degree. Both pollution levels and property values differed in the three cities, of course, but in all three a moderate decrease in air quality—5 percent to 15 percent—correlated with a significant decrease in property values—from $300 to $700 for an average property. Roughly each 1 percent increase in either sulphation or particulates was associated with a .08 percent decline in the price or the rental value.

With these results, it is possible to calculate a damage figure for any city in the U.S. if certain information is available: median property value the number of owner-occupied residential properties, and indices of su phur trioxide and particulate pollution. To determine *annual* damage the analyst multiplies the lost value by an interest rate to account fo what the lost value would have earned if pollution hadn't wiped it ou In residential real estate a 12 percent rate of discount seems appropriate Such calculations were carried out for eighty-five U.S. cities for 196 (more recent air-sampling data were not available). The combined prop erty-value losses worked out to $621 million.

A DECISIVE CASE FOR CONTROL

To complete the benefit-cost analysis, we need figures for the costs o canceling these damages. The National Air Pollution Control Administra tion has worked out some figures for the costs of controlling sulphur ox ides and particulates in those eighty-five cities. The levels of control cor respond to improvements in air quality of from 5 percent to 15 percent according to Anderson, who helped prepare the NAPCA study. For fisca 1972 (the first year for which NAPCA calculated a full set of cost figures the low estimate of total costs came to $609 million.

So we have figures for 1965 damages and fiscal 1972 costs. To match them up, we must either deflate costs to 1965 prices or inflate 1965 prop erty values to their projected 1972 levels. With either calculation, th marginal benefits of air-pollution control would far exceed the margina costs. This conclusion, moreover, obviously understates the economic cas for air-pollution abatement. Losses in residential property values encom pass only a part, and perhaps not even a major art, of total air-pollutio damages. Anything like a complete benefit-cost analysis—which woul be an enormous undertaking—would clearly result in a decisive economi case for air-pollution control.

In a sense the water researcher faces a more formidable problem o damage estimation than Crocker and Anderson did. Property-value losse can be readily translated into dollar magnitudes, but a major benefit o water-pollution control would be more frequent and more satisfying out door recreational experiences—a commodity whose dollar value is esti mated only with great difficulty. The researcher, in effect, has to assig what economists call a "shadow price" to a day's use of water recreatio facilities. One way of doing this is to find out what the recreationist i paying for comparable services at private facilities. Another way is to us

, method that is based on the cost of traveling to the public recreational site.

If the researcher can multiply a reasonable shadow price by the increase in recreational activity that would result from a given improvement in water quality, he can chart a curve of marginal benefits. If he can also determine what it would cost to achieve that improvement in water quality, he has a marginal cost curve. The point of intersection of the two curves represents the optimal improvement in water quality from the viewpoint of benefit-cost analysis.

WHAT'S A CLEAN RIVER WORTH?

This kind of study has been done for the Delaware River by economists Paul Davidson, F. Gerard Adams, and Joseph Seneca. What they considered on the benefit side was increased recreational use of the river for boating and fishing, specifically over the period 1965–90. On the other side were the costs of specified improvements in water quality—as measured by levels of dissolved oxygen—to make the river more suitable for boating and fishing. The three economists found that it took only very moderate shadow prices for recreation to justify the costs of the improvements. For example, it would have paid to clean up the Delaware considerably in 1965 if the use of the river for one day's boating was worth as much as $2.55 to the boater.

The use of such a price in benefit-cost calculations does not mean that the users of the facilities would be required to pay it, though that is conceivable. The point is that if such a shadow price is reasonable, enough additional social welfare (shadow price times extra recreational use) will be generated over the twenty-five-year span to repay the investment in river quality.

Some economists find fault with the Delaware River study on the ground that too low an interest rate—5 percent— was used in discounting future benefits. While this rate is higher than the 3 to 4 percent typically used for public projects, it is much below realistic rates for the private economy, even in 1965. The use of 5 percent means, in effect, that funds worth, say, 10 percent in the private sector would be diverted and put to work in public projects at about one-half the yield. In straight economic terms this is plainly inefficient. Says George S. Tolley, professor of economics at the University of Chicago: "I believe in letting the chips fall where they may. If a public project cannot stand on its own it should not be buttressed by artificially low discount rates. In fact, if proper discount rates had been used, many of the dams and reservoirs built over the past few decades would never have been started. They couldn't have paid their way."

If the three economists had used a more "realistic" discount rate, they would have had to raise the shadow price for boating in order to justify cleaning up the Delaware to the specified levels of water quality. Since $2.55 was probably a bargain for a day's boating even in 1965, they have some room to maneuver. What's more, increased recreational use is obviously not the only benefit from a cleaner Delaware River. When a river is upgraded, riparian property values rise dramatically. In proper benefit-cost accounting, each such subsidiary benefit should be taken into account.

THE THIRD POLLUTION

Recreational benefits and higher property values are also important in benefit-cost analysis of solid wastes—garbage, trash, ashes, sewage sludge, building rubble, auto hulks, beer cans. Sometimes called the "third pollution," solid wastes are normally regarded as a nuisance, or worse. They can also be regarded as resources—resources out of place. The general objective of solid-waste control might be thought of as the displacement of wastes from locations where they have negative value to locations where they have positive value.

One means of doing that is the sanitary landfill. Untreated waste is buried daily in layers, each covered under several inches of compacted earth. The landfill technique has transformed thousands of acres of low-value land into parks, playgrounds, golf courses, bathing beaches, marinas, parking areas, and other useful facilities. While it is sometimes said that the future of the landfill is limited by land scarcity, there is actually plenty of low-value land to be improved. Karl Wolf, head of research of the American Public Works Association, points out that strip-mining operations mangle more than 150,000 acres each year. If this ruined land were used for landfill, it might be enough to hold our entire annual output of solid wastes.

In areas where the landfill is surrounded by residential or commercial property, transitional problems arise. Although completed landfills improve property values, the actual filling operations temporarily impair values. People living near an active landfill have to contend with refuse-truck traffic, noise, odor, and unpleasant views. The longer these disamenities persist, the more basis property owners have for demanding compensation. Unfortunately, large landfills may take quite a few years to complete. And community landfills tend to be large, partly because they tir resentment; politicians, wanting to limit the number of angry voters, refer one big site to several smaller sites.

One way to speed up completion of large landfills would be to pool wastes over extensive areas. This approach, of course, would require the cooperation of independent municipalities. It would also require some inexpensive means of long-distance haulage. The railroad seems the likeliest answer here. Since many different transport operations are required for large landfills—truck collection of waste, transfer to train, retransfer to trucks, and dumping at the landfill site—it may even make sense for the railroads to become vertically integrated waste disposers.

Landfill can be thought of as a form of recycling—in this case, using wastes as a construction material. Wastes may have higher value as fuel, and some wastes still higher value as raw materials. Quite a bit of trash and garbage is incinerated in the U.S. today, but in ways that change wastes from one form to a possibly even less desirable form, air pollutants. It is possible to design efficient incineration systems that emit little pollution and yield useful heat. In Europe utilities burn trash to provide steam, but this use of solid wastes is relatively rare in the U.S. It might be more economically attractive if refuse vehicles were designed to pick up tanks of used oil from service stations and spray their contents on household garbage as it is collected, thereby raising the BTU content of refuse.

Because there is a pervasive throwaway psychology in the U.S., we do not come close to realizing—or even envisioning—the potentialities of recycling. In many instances where recycling is dismissed as economically or technically unfeasible, the possibility has not been carefully examined. The steel industry nowadays recycles much less scrap than it used to because the basic oxygen furnace, unlike the older open hearth, supposedly cannot assimilate very much cold scrap; industry leaders argue that heating costs make recycling uneconomic. Research scientists at M.I.T. disagree. They point out that the basic oxygen furnace produces a lot of hot waste gases. These could be passed up a tower filled with pulverized steel scrap, which could thus be heated enough for recycling.

THE BEER-CAN PROBLEM

An increasingly troublesome category of solid wastes in the U.S. is the detritus of what has been called "the packaging explosion." In packaging, we have moved away from recycling instead of toward it. The returnable beverage bottle of yesteryear, a casualty of affluence, has given way to the throwaway bottle and the throwaway can. The can, what's more, is not rustable steel anymore but persistent aluminum. In an affluent democracy, it is not feasible either to make people stop littering the landscape with beer cans and similar artifacts or to pay other people to go around picking them up. But it should be possible to arrange matters so that a lot more used containers get recycled. Two problems are involved here—to induce people to return containers to the retailer and to make it worth the retailer's while to bother with the things.

The prospect of getting back a small deposit is for most people these days a weak inducement. The popularity of lotteries suggests a more effective alternative: some kind of arrangement by which the return of every five-hundredth or thousandth soft-drink bottle or beer can pays the lucky returner a windfall of, say, $5 or $10. Where would the money come from? Partly from the recycle value of the containers and partly from taxes applied by society at appropriate points.

One kind of tax that might be imposed is a levy on containers, graduated according to the difficulty of recycling. There always seems to be some perverse component that gets in the way of economical recycling. The aluminum ring that the twist-off cap leaves around the neck of the bottle makes it uneconomic to grind such bottles into cullet for glassmaking; the cost of removing the metal, either before or after grinding, is too high. The small magnesium content of the aluminum can lowers its salvage value. And the tin coating and lead solder on the otherwise steel "tin can" largely exclude it from the salvage market. Unless this obstructive heterogeneity can be dealt with, society is unlikely to make much of a dent in what might be called "the beer-can problem." If containers do not have a worthwhile salvage value, the retailer will balk at assuming the costs involved in handling them—they would become *his* waste-disposal problem.

In some measure, government can help alleviate the problem through discriminatory procurement policies. Federal purchases account for about 6 percent of total packaging expenditures. If government were to insist upon tin-free steel cans or magnesium-free aluminum cans, industry would be powerfully prodded to alter the technology. Taxation, however, would provide a more comprehensive approach. To be effective, such a tax would have to be high enough to provide those lottery-type payoffs *plus* subsidies to cover the extra costs of putting the salvaged material into recyclable forms. If the tax were well designed, it would encourage the manufacture of more recyclable containers. Ideally, the design of the tax should also encourage degradability—not all containers would be recycled in any case, and it is desirable that litter be as evanescent as possible. Some cheerful scientists even look forward to the "biodegradable beer can," which would hold together on the grocer's shelf but succumb to bacterial decay after it was thrown away.

O EACH HIS OWN LEAST-COST MIX

The disposal tax seems to be on the right track in that it makes the pol-
uter assume costs related to his pollution. In the water and air pollution
ields, proposals similar to the disposal tax are usually called "effluent
charges." The concept of the effluent charge is beginning to get a friendly
hearing in Congress. Last November, Senator William Proxmire of Wisconsin introduced a bill to place an effluent charge on water polluters. Al-
hough no figures are presented in the bill, Resources for the Future,
which helped Proxmire prepare the legislation, has suggested a charge of
8 to 10 cents per pound of biochemical oxygen demand (BOD) added to
any waterway. If a levy of this kind were imposed nationally—and some
hink companies might relocate if it weren't—national revenues at the 8 to
10 cent level would come to around $2 billion to $3 billion. Part of the
revenues could go to finance regional development authorities that would
undertake research and plan and construct collective treatment facilities.

Advocates of the effluent-charge approach argue that it would operate far more economically than the present arrangements for water-pollution abatement. Currently the federal government attempts to set—or get
states to set—waste-treatment standards requiring large industrial and municipal expenditures, and provides some financial assistance, generally to
municipalities. These standards often call for excessive uniformity, in
hat every producer in a given river basin must abate by the same percentage. Such a program is bound to be inefficient. For one thing, it requires small plants with high marginal cleanup costs to treat to the same
level as large plants, which usually have lower marginal costs. For another, it ignores locational factors and stream hydrology—i.e., the obvious
circumstance that a polluter in one part of a stream can discharge wastes
without causing any damage, while another polluter elsewhere can cause
a lot of damage with the same amount of wastes.

The effluent charge remedies at least the first deficiency, because
each water user along the stream can find his own least-cost mix of pollu-
tion-abatement measures and effluent-charge payments. Moreover, an adequately designed system of effluent charges gives the polluter an incentive to carry treatment further than he would under a uniform standard.
If the standard is 85 percent removal of biochemical oxygen demand, the
polluter will presumably not exceed it. But if he has to pay for all BOD,
he has an incentive to push treatment to a higher level, say 95 percent
removal, as long as the additional cost is less than the effluent charges
avoided.

The superiority of the effluent charge over the uniform standard

has been documented in a 1967 study of the Delaware estuary by Gran
W. Schaumburg of Harvard. He found that if proportional abatemen
were required in the Delaware, the area's forty-four major polluters-
industrial and municipal—would pay a total of $106 million to clean u
the water to a specified level of dissolved oxygen; but under a reasonabl
system of effluent charges, the combined cost would be only $61 millior
Further cost reduction could probably be achieved by placing entir
river basins under the jurisdiction of regional authorities that would buil
and administer large-scale treatment facilities. These can yield substantia
economies of scale. Economists Andrew Whinston and Glenn Grave
reckon that if four polluters on the Delaware, each with a waterflow o
2,500,000 gallons a day, built individual treatment plants to remove 5(
percent of BOD, the cost of all four together would come to $1,520,00(
a year. But if the four were located close enough to each other, the
could become partners in a single plant and save a total of $480,000
year.

DOWN, REPLIED THE COMPUTER

Whether Congress selects the effluent-charge route or not, it seems clea
that the public is of a mind to devote more of the economy's resources t
environmental improvement. It is impossible to predict with confidenc
how much the coming campaign to reverse environmental deterioratio
will cost, or what the impact will be on the national economic accounts
With the aid of an econometric model, however, it is possible to explor
this largely uncharted territory.

Econometric models of the U.S. economy can be used to estimat
the effects of an altered investment mix resulting from the imposition o
additional pollution controls. Choosing a model that had predictec
changes in the U.S. economy with reasonable accuracy during 1962–64
economist Robert Anderson altered some investment and price inputs tc
reflect a fairly stringent level of air-pollution control. He then reran the
model for the two-and-a-half-year span from the first quarter of 196:
through the second quarter of 1964. Anderson assumed that (1) manu
facturing industries were required to increase their investment on air
pollution devices at an annual rate of $1.2 billion; (2) public utilities wer
required to increase *their* outlays at an annual rate of $320 million; anc
(3) new-car prices rose by 1 percent owing to the installation of emission
control devices.

What happened to G.N.P.? It went down. At the end of the twc
and a half years, that is, G.N.P. was at an annual rate of $617 billion
whereas without the assumed pollution controls the model had predictec

1ore than $625 billion. Unemployment was up: the new rate worked out
) 5.3 percent instead of 4.8. And prices were up too—the price level was
bout 1.2 percentage points higher than had been projected.

For the first four quarters of the simulation, G.N.P. reached higher
:vels with the air-pollution controls than it would have without them—
ut largely because of price effects. In the very short run, demand for
1ost goods tends to be "price inelastic"—that is, most people don't change
1eir buying habits quickly as prices rise. As a result, industry's increased
osts of coping with air pollution could temporarily be passed on to con-
umers in the form of higher prices, which are, of course, reflected in
;.N.P. But after a while the decline in disposable income tends to stiffen
esistance to price increases. Demand weakens and output starts going
own.

>NCE SOCIETY CHARGES RENT

"his exercise in retroactive simulation suggests that increasing expendi-
ures on pollution control would produce some of the effects of a reces-
ion. But for this kind of assessment, econometric models are biased on
he side of pessimism. For one thing, the model is incapable of taking
echnological change into account. In this case, that is an especially seri-
1us shortcoming. The spectrums of technological options for pollution
ontrol are likely to expand quite dramatically, even in the short run.
\batement technology is just in its infancy. Since society has for so long
llowed free-of-charge use of the natural environment for waste disposal,
ncentives to do research in abatement have been weak at best. As a re-
ult, pollution-control equipment tends to be relatively inefficient, and
he potential trade-offs among waste treatment, process changes, and re-
ycling are poorly understood. Once society—by one means or another—
)egins charging rent for use of the environment's capacity to absorb
vastes, engineers will have to think about pollution control as an integral
)art of plant design rather than as an afterthought. A lot more research
unds will be allocated to pollution control, and costs may go down faster
han anyone expects.

It is even conceivable that industrial costs will not rise at all in the
nedium or longer term. Pollution control not only provides incentives for
nore efficient operation through recycling, but also makes cities more
ivable. And people who work in more livable places don't have to be
)aid quite so much as those who work in less livable places. If wage rates
n the future are just slightly lower than they would have been if the
:ities had remained polluted, the difference might quickly offset industry's
ncreased pollution-control costs.

Over the longer term, pollution abatement seems likely to *increase* real G.N.P. A significant decrease in air pollution, for example, can be expected to reduce absenteeism and turnover and improve productivity. Some industries, perhaps many industries, might have to pay out less in sickness and death benefits. With turnover reduced, they might also have lower training costs. If these longer-range savings were put into a thorough benefit-cost analysis, many corporations might discover that pollution control yields a profit, entirely apart from any altruistic considerations.

THE GROSS NATIONAL EFFLUENT

In a broader sense it is a mistake to put any great emphasis on the G.N.P. aspect. Although the costs of environmental improvement are reflected in national product, many benefits are not. For example, when property values rise because of a decline in air pollution, the community's real wealth or capital stock increases; but this shows up in G.N.P. only to the extent that actual or imputed rents go up or real-estate salesmen's commissions get bigger. Similarly, although environmental improvement may enrich leisure and so increase satisfactions (or reduce dissatisfactions), these benefits could not be reflected in G.N.P. at all, as G.N.P. is presently reckoned.

Moreover, it is difficult to know what is really "net" in gross national product since the measurement counts both output and some of its adverse consequences—that is, every good and service that has a price. To arrive at even an approximate gauge of the annual increase (if any) in welfare, we would have to deflate G.N.P. by a magnitude that might be called G.N.E.—gross national effluent. G.N.E. would be a statistical basket for all those negative goods and services produced in the course of, or as a result of, the production of positive goods and services. Negative goods and services in this sense include additional transportation to escape the effects of environmental impairments, additional cleaning, additional medical services, goods prematurely replaced because of soiling or corrosion, and, of course, pollution-control equipment. If we subtracted G.N.E. from G.N.P., the remainder would be a better measure than G.N.P. of what the economy has done for us in any year. And it is certain that pollution control will sharply increase the value of this remainder.

PART 3

CURRENT CASES

Environmental disruption in Japan: again the Japanese outdo us

As a country with perhaps the world's highest and most sustained rate of economic growth, Japan is a fascinating study not only of rapid industrialization but also of the environmental disruption that results when modernization comes too fast and haphazardly. The editor prepared this article after having attended the International Symposium on Environmental Disruption in the Modern World—A Challenge to Social Scientists held in Japan in March 1970. The symposium was sponsored by the Standing Committee on Environmental Disruption of the International Social Science Council.

Squeeze half the American population into the state of Maryland and you have some idea of the problem Japanese ecologists and economists have to contend with today.[1] Their task is further complicated by the fact that their economy has been growing so rapidly that no one has had the time to devote enough attention to the disposal of wastes that a 10–12 percent annual increase in GNP can generate.

To the visiting ecologist, Japan suggests what might happen the day before the earth poisons itself to death. It almost seems as if the Japanese had agreed to conduct a laboratory experiment to demonstrate the kinds of ecological horrors man is capable of inflicting upon himself without having to go to war. For example, although the rest of the world has suddenly become preoccupied with the threat of mercury poisoning, the first recorded case in Japan occurred in the town of Minamata as early as 1953. Since that time, 116 Japanese citizens have been paralyzed as a result of the "Minamata Disease" (mercury poisoning), including forty-

[1]For a more thorough analysis of environmental disruption in Japan, from which most of the material for this article is drawn, see the essay by Shigeto Tsuru in Shigeto Tsuru, ed., *Proceedings of International Symposium on Environmental Disruption*, Tokyo, Asahi Evening News, March 1970.

five who died. A second outbreak of the Minamata Disease occurred in 1965 in Niigata, where an additional thirty-nine people were stricken and six later died.

The Japanese suffer from perverse environmental maladies that we have not even worried about yet. As early as 1955, doctors in Toyama prefecture were puzzled by what came to be called "Itai, Itai (Ouch Ouch!) disease," because it was so painful. Eventually fifty-six people died from what apparently was the discharge of cadmium into the water. Like mercury, cadmium builds up rapidly in the food chain. Hundreds also suffer yearly from "Yokkaichi asthma," which results from a severe form of air pollution. These are just some of the more extreme forms of misery which stem from misuse of the environment. Why is it that the Japanese, who have always treated nature so delicately and reverently have suddenly become so destructive and abusive toward their environment?

The underlying factors are not much different from those in other industrialized countries like the United States and the Soviet Union. It is just that they have all exploded so much faster and with a much greater intensity in the crowded Japanese islands. After World War II the glorification of an ever larger GNP formed the basis of a new materialism which became a sacred obligation for all Japanese governments, businesses, and trade unions. Anyone who mentioned the undesirable byproducts of rapid economic growth was treated as a heretic. Consequently everything possible was done to make conditions easy for the manufacturers. Few dared question the wisdom of discharging untreated waste into the nearest water body or untreated smoke into the atmosphere. This silence was maintained by union leaders as well as most of the country's radicals; except for a few isolated voices, no one protested. An insistence on treatment of the various effluents would have necessitated expenditures on treatment equipment that in turn would have given rise to higher operating costs. Obviously this would have meant higher prices for Japanese goods, and ultimately fewer sales and lower industrial growth and GNP.

The pursuit of nothing but economic growth is illustrated by the response of the Japanese government to the American educational mission that visited Japan in 1947. After surveying Japan's educational program, the Americans suggested that the Japanese fill in their curriculum gap by creating departments in chemical and sanitary engineering. Immediately, chemical engineering departments were established in all the country's universities and technical institutes. In contrast, the recommendation to form sanitary engineering departments was more or less ignored, because they could bring no profit. By 1960, only two second-

ate universities, Kyoto and Hokkaido, were interested enough to open
uch departments.

The reluctance to divert funds from production to conservation is
xplanation enough for a certain degree of pollution, but the situation
vas made worse by the type of technology the Japanese chose to adopt
or their industrial expansion. For the most part, they simply copied
American industrial methods. This meant that methods originally de-
igned for use in a country that stretched from the Atlantic to the Pacific
vith lots of air and water to use as sewage receptacles were adopted for
an area a fraction of the size. Moreover the Japanese diet was much more
lependent on water as a source of fish and as an input in the irrigation of
ice; consequently discharged wastes built up much more rapidly in the
ood chain.

These difficulties have been intensified as the Japanese give increas-
ng emphasis to the production of synthetic products. The production of
lacron and nylon, for example, normally necessitates the use of considerably more water than the processing of a similar amount of wool
and cotton.

Japanese inter-governmental relationships also complicate efforts to
alleviate the damage. Almost all governmental units below the federal
evel are denied any significant taxing power. Municipalities are required
o transfer 70–90 percent of their tax revenue to the national government.
Consequently, even if they want to improve conditions, local governmental authorities must seek the help of the national authorities who
eretofore have been unwilling to allocate funds even for sanitation
control. Thus only 36 percent of all the homes in Tokyo, the most heavily
populated city in the world, are connected to sewers.

Until recently the cities and provinces did not even have the power
o prevent small factories from drilling for water in the center of urban
areas. Since 1925 this has given rise to what the Japanese consider
a new form of environmental disruption. As a result of this drilling, the
water table has subsided along with the land which rests upon it. Although private drilling for water was banned in 1966, the situation has
not markedly improved—some suspect that the drilling goes on surrepti-
tiously. Today about five square miles of land in the east end of Tokyo
lie more than six feet below sea level. Osaka has a similar problem. In
both cities, hundreds of thousands of people live and work in areas that
are now below sea level, protected only by fragile dikes. There is a des-
perate fear that the next typhoon that hits these cities will drown thousands of inhabitants.

The disruption of the environment, or *kogai*, as the Japanese call it,
has become especially intense within the last decade. The linkage be-

tween industrialization and environmental disruption is perhaps best illustrated by what happened in the city of Yokkaichi, the home of Yok kaichi asthma. Until 1955, Yokkaichi was a quiet seaside resort noted fo its clean air and water and good fish. About the only sign of industria activity was the remains of an old Japanese naval oil depot that had bee destroyed by the United States in World War II. Recognizing the valu of oil over coal as a source of fuel, Mitsubishi and Shell Oil formed partnership in the mid 1950s to build an oil refinery to supply Japanes industrial needs. Regrettably, the refinery, which was completed in 1958 was supplied with sour Middle East oil, which is very high in sulfur. landfill development was extended into the bay and a petrochemica complex and thermal electric power plant was built at the edge of th city alongside private seaside homes and, of all things, a hospital. Pro duction in Yokkaichi soon began to grow twice as fast as that of the res of the country, which already exceeded most industrialized areas of th world. The first reaction to this unrestrained economic explosion too place in 1959, when a noticeable increase in asthma occurred and th death rate increased sharply. Patients suffering from what came to b called Yokkaichi asthma were taken naturally enough, to the hospital located in the middle of all this noxious activity. A second reaction wa that shortly thereafter, the fish caught in the bay developed an unpleas ant odor from the effluent dumped into the water, which made it impos sible for them to be sold.

The rapid industrialization was accompanied by a similar rapid ur banization. The rural population, which constituted 48 percent of th population in 1930, fell to 20 percent in 1970. The rest of the populatio has crowded into the cities, bringing additional wastes. Moreover as th national and personal wealth of the country has grown, consumption ha increased at an even faster rate. This means not only more solid an liquid wastes waiting for disposal but also an increase in congestion an combustion—inevitable by-products resulting from the increase of au tomobile stock from twenty-three thousand cars in 1947 to two millio in 1970. Moreover even though Japanese cars intended for the America market had been equipped with exhaust controls for some time, no suc equipment was required on Japanese cars produced for the domesti market until late 1970. An indication of the rapidity with which *koga* encompassed the country is the fact that the word itself was not even i the Japanese dictionaries of 1955.

It was not until 1970 that Premier Sato's government agreed to con front the problem of pollution. This was largely in response to a shar deterioration in environmental conditions and to the protests whic eventually did take place. When opposition did come, significantl

nough, it was led by groups of farmers and fishermen who sometimes
esorted to violence out of frustration and wrath when their crops and
atches were destroyed or contaminated by industrial wastes. This was
'ue at Minamata and Yokkaichi. Later the ranks of the farmers and
shermen were joined by women's and consumer groups. Almost all the
ther segments of Japanese society were reluctant to do anything that
'ould jeopardize their jobs or hamper Japanese industrial efforts. Only
roups outside the industrial grasp could risk such protests. But even
hough the Japanese government has now vowed to curb environmental
isruption, there is considerable doubt as to just how determined it is to
ismantle or filter the chimneys and pipes that are laying the golden but
olluted eggs of industrial growth. Some critics fear that Japanese indus-
'y may try to solve its difficulties merely by relocating in the developing
ountries of southeast Asia. By sweeping the effluent under someone else's
nvironment the Japanese may be better off, but not the world.

The following articles consider the success and failure of pollution control in a selected number of river and air basins. There is much that is impressive and much that is discouraging.

Water quality management by regional authorities in the Ruhr area

ALLEN V. KNEESE

Allen V. Kneese is an economist on the staff of Resources for the Future and the author of The Economics of Regional Water Quality Management, *the foremost study of the economics of water pollution. In this article he describes how pollution has been successfully checked by a method which combines regional cooperation and management with a system of economic charges to polluters. As a result economic incentives are used to control the pollution of manufacturers and municipalities.*

PART I—BACKGROUND AND CONCEPTS

Water quality management is coming to dominate the problem of planning for development and use of water resources in many parts of the United States. Moreover, it has become widely recognized that water quality is a problem which, in most respects, can be best analyzed and dealt with on a regional basis. This is seen in the creation of the Delaware River Basin Commission, one major function of which will be the management of water quality, the United States Public Health Service

Reprinted by permission from *Papers and Proceedings of the Regional Science Association,* Volume 11, 1963.

Comprehensive Planning Studies for the various river basins and the establishment of numerous watershed and metropolitan authorities.

The recent report of the Senate Select Committee on National Water Resources helps give perspective to the possible magnitude of the water quality management task in the various water resource regions. Indeed it replaces the prevalent image of quantitative shortage on a nationwide scale with the conclusion that in most areas water supply is more than sufficient to meet the various projected uses which man will make of concentrated supplies. But it also concludes that presently dependable supplies are generally far from adequate to provide dilutions of projected future municipal and industrial waste discharge. Based on its analysis, the committee concluded that maintaining comparatively clean streams in the various regions might require a national investment of an additional $100 billion by 2000 A.D. This is indeed a huge sum. By contrast the cost of completing the Bureau of Reclamation program of multi-purpose western water resources development is estimated at a mere $4 billion after 1954.

The estimates of the committee cannot be viewed as more than broad indicators of the potential magnitude of various aspects of water supply problems. *They do suggest, however, that achieving fairly clean streams throughout the United States, in view of expected future economic-demographic development, will involve public investment far higher than in any other field of resources development or conservation.*

Despite the fact that some institutions have been created whose major or primary function is to plan for optimum water quality management on a regional basis, there is as yet little institutional, economic or engineering-scientific analysis aimed explicitly at the problems of regional water quality management in the watersheds and river basins of the United States.

For example, questions concerning the appropriate spatial jurisdiction and appropriate powers of an authority responsible for water quality in a region have hardly been addressed. There are many such questions. For example, should a single authority be made responsible for the water resources of a small watershed, a river basin or a whole system of river basins? If there are different authorities, what coordinating devices are available? Should authorities only make general framework plans or should they be directly responsible for their execution by (say) constructing and operating facilities (dams, treatment plants, etc.)? How should the obvious interdependencies between land use patterns (especially industrial and municipal location decisions) and water quality problems be handled? Where land use planning is undertaken should it be under the same authority as water resources? If not, what functional division of powers is appropriate? What information should the authorities provid each other? What incentive or disincentive devices should they hav available to see to it that social costs which occur in their area of respon sibility are reflected in decisions of fiscally independent decision makers

In light of increasingly pressing questions of this kind, it is useful t examine the experience of an area where these problems have long bee confronted and regional institutions have been developed for dealing wit them. The area in question is the Ruhr industrial area of West Germany Indeed, the Ruhr with its extremely concentrated economic and demo graphic development and with the extraordinarily heavy burden which it urban-industrial area-society puts on water resources, reflects the type o situation numerous areas in the United States are just beginning to face

The aim of this paper is to provide a brief review and assessment o the regional quality management activities of the water authorities in th Ruhr area. Special emphasis is placed upon the methods used to articu late the system planning and operation activities directly under the control, with often equally important decisions impinging upon wate quality but under the control of other private and public or semipubli decision makers. It will be seen that the cost assessment and distributio methods used have played a prominent role in this regard.

Before undertaking a discussion of water quality planning in th Ruhr area, I would like to make a few generalizations concerning th economics of resource allocation especially pertinent to water qualit management. This will provide a setting for the subsequent discussio Economists will readily understand that unrestricted waste disposal int "common" water courses produces technological external diseconomie Since putting wastes into water courses gives rise to costs which occu primarily "offsite," the cost and production structure tends to be distortec There is no inducement to undertake waste water treatment, and othe abatement measures, materials recovery, process adjustments and othe measures to reduce waste loads generated are not implemented to an opt mum degree. The effectiveness of process engineering and materials re covery processes in reducing waste loads has been richly demonstrate by various instances in this country and perhaps even more strikingly i the Ruhr area. Moreover, industry accounts for over two-thirds of th organic waste load in the United States and a far higher proportion o most other pollutants. This emphasizes the importance of regional wate resources authorities providing the appropriate incentives for reductio of industrial waste loads by in some fashion causing offsite costs to b reflected in waste generation and disposal decisions.

Indeed the traditional tax subsidy solution of welfare economics t externality problems of this kind lays exclusive emphasis upon the incer ves provided by means of a redistribution of opportunity costs. I have lsewhere reviewed the circumstances under which this suggested solu-on might work in practice and some approaches to the complex meas-rement and computational problems that are encountered, including ıch matters as difficult to value as damages, interrelationships between ·astes, hydrological variability, etc. The general principle is, however, ıat an effluent charge equal to downstream damages (resulting from in-·eased water supply treatment and value of physical damages) would be ‌nposed on the decision unit responsible for the outfall. The decision ‌nit would then take measures to reduce its waste discharge by an opti-‌ıum combination of measures—process and product adjustments, waste ·eatment and perhaps others, until the marginal cost of an additional ‌nit of abatement equals the marginal cost of damages imposed in the ffected region. At this point the cost associated with the disposal of ·astes in the region would be minimized. Several things may be noted ‌vithout elaboration here) about this solution with a view to the litera-‌ıre on external diseconomies:

1. Since "water floweth whither it listeth," the reciprocal effects ·hich have recently come under discussion and which may destroy the ‌ossibility of achieving approximately optimum results via the charges or ‌ıxes route are not involved, at least in the case of streams. Pollution ‌amage is essentially a serial phenomenon.

2. In order to obtain an efficient (cost minimizing) solution, the ‌amaged parties need not be compensated. If the charge levied upon the ·aste discharger produces optimal water quality in the stream, the water ‌sers will be induced to adapt optimally to it. (This result may not be ‌onsidered equitable but this paper offers no criteria of equity.)

3. Solely from the viewpoint of efficiency, the desired (cost mini-‌ıizing) result can be achieved with a subsidy (for reducing effluent) ‌ased on damages. The subsidy must be paid to the waste dischargers ‌ince, for the reason indicated in (1), only they control the water quality ‌vailable to downstream users. The significant thing is that the upstream ‌·aste discharger must view downstream costs as opportunity costs. The ‌resent discussion will, however, proceed in terms of effluent charges.

4. Effluent charges have not been used on a regional basis in this ‌ountry. In some instances effluent "standards" have been proposed and ‌sed. Charges appear to have certain advantages over standards which ‌annot be examined here. Both, however, can be viewed as devices for ‌edistributing opportunity costs with attendant effects on treatment, proc-‌ss and product adjustments, and industrial location decisions.

5. If there are economies of scale in abatement measures, whic cannot be realized by individual waste dischargers, the "classical" ta approach (or analogous procedures) cannot achieve an optimal solution.

The last point requires some additional comment since it is basic t further discussion of the work of the German water authorities. Whe economies of scale exist in abatement measures "system design" arises a a problem confronting a regional authority, in addition to seeing to that opportunity costs are distributed in such a way as to induce efficier behavior. This is true, for example, when economies of large scale exis in waste treatment which permit wastes from diverse sources to b brought together for collective treatment, or where measures such a augmentation of low steamflows by reservoir releases or artificial reaera tion of streams are efficient alternatives or supplements to treatment ove certain ranges, or where the entire flow of a stream can be advanta geously treated. In other words, the problem of system design presents it self when economical abatement measures exist which cannot be undei taken by individual polluters.

In virtually all highly developed regions efficient ways of contro] ling pollution will be available, the use of which cannot be induced b levying the net offsite costs of their waste disposal on the waste disposer: It is to such regions that the present paper is especially relevant.

In these areas a social cost-minimizing solution will demand plan ning of the system by an organization which can comprehend the signi icant large-scale alternatives and supervise their operation. Such an or ganization is thus confronted with a set of problems. These includ designing the system, operating some elements of it and making charge (or using other devices) to induce efficient use of alternatives, the con struction and operation of which it does not control directly. The latte would ordinarily, and probably should, include *at least* process and pro duction adjustments by manufacturers' pretreatment of wastes, and deci sions with respect to specific locations of industries.

If the objective of the regional authority is to minimize the sum c the costs associated with waste disposal in the region, the economic de sign criterion would be to equalize the costs of all alternatives *includin, the costs of pollution damages* at the margin. This would be accomplishe by a combination of direct construction and operation of abatemen measures and by effluent standards or charges. The latter can be show] to lead to more efficient industry responses than arbitrarily selected stand ards, and are especially advantageous in inducing adjustment to shortrur changes in social cost. The abatement facilities actually constructed an operated by the authority must take into account the response of th waste load generated to the system of charges. The optimal mix betwee]

easures planned and executed by the regional authority and those in-
uced on the part of other decision makers will of course depend upon
ıe degree to which economies of collective measures can be realized.
his in turn will largely depend upon the extent of development of the
asin.

One of a number of difficulties in actually designing and operating
system which minimizes costs is that certain values diminished or de-
royed by water pollution are exceedingly difficult to measure. Promi-
ent among these are the value of aesthetic and recreational amenities.
Vhere these values are not actually quantified, specific judgments con-
erning their physical requirements or standards can be made. In a
ormal sense they can be considered constraints on the objective (which
'e have taken to be the minimization of costs associated with waste dis-
osal). For example, costs may be minimized provided that dissolved
xygen does not fall below four parts per million (ppm) in the stream (a
enerally accepted minimum level for fish life). There are various formal
ıethods by which such constrained optimum problems can be solved. In
eneral, the economic criteria for an optimum with such constraints are
ıalogous to those indicated earlier (*i.e.* the equalization of marginal costs
ı all directions). In other words, the optimum system is not attained
ntil a situation is reached in which it is impossible to make marginal
radeoffs" (say, between waste treatment and pollution damages) which
)wer costs without violating the constraints. The marginal costs affected
y the constraint (say, waste dilution) now, however, have a shadow
rice which derives from the limited supply of the constrained input (say,
issolved oxygen).

These general points have been made not to imply that *precisely*
ptimum (given the objective) regional water quality management sys-
?ms are possible or even desirable goals but to provide a conceptual
amework for discussion of the German experience.

ART II—THE GENOSSENSCHAFTEN

'here are seven large water resources Cooperative Associations called
:enossenschaften in the highly industrialized and heavily populated area
enerally known as the Ruhr. These organizations were created by spe-
ial legislation in the period from 1904 to 1958.[1] There are thousands of

---

[1]All of the Genossenschaften with one exception were established before 1930.
he Erftverband (Verband and Genossenschaft are used interchangeably in this con-
:xt) was created in 1958 primarily to deal with problems resulting from a massive
ımping down of ground water tables by the coal industry in the area of Erft river,
est of the Rhine.

water Genossenschaften in Germany, most of them created for specia
purposes such as the drainage or flood protection of specific plots of land
The large Genossenschaften in the Ruhr region were given almost com
plete multipurpose authority over water quantity and quality in entir
watersheds by their special laws. These organizations are henceforth re
ferred to simply as *the Genossenschaften.* They have for up to fifty year
made comprehensive plans for waste disposal, water supply, flood contro
and land drainage (a problem of great significance in the coal minin
areas). The Genossenschaften are comparable to cooperatives (in th
Anglo-American sense) but with voting power distributed in accordanc
with the size of the contribution made to the associations' expenditure
and with compulsory membership. Members of the associations are prir
cipally the municipal and rural administrative districts, coal mines an
industrial enterprises.

General public supervision over the Genossenschaften is in the hand
of the Ministry of Food, Agriculture and Forestry, of the state of North
Rhine Westphalia. The Ministry's supervision is, however, almost com
pletely limited to seeing that the associations comply with the provision
of their constitutions.

The Genossenschaften have the authority to plan and construc
facilities for water resources management and to assess their member
with the cost of constructing and operating such facilities. A process c
appeal (internal to special boards and final appeal to the federal admir
istrative courts) is available to the individual members.

Statutes creating the Genossenschaften are limited to a few brie
pages. Accordingly, the goals and responsibilities of these organization
are set forth in highly general terms. This has left the staffs and th
members free to adapt to changing conditions and to develop procedure
and concepts in line with experience. Perhaps one general provision c
the statutes, the meaning of which has developed greater and greate
specificity in the course of time, has proved most central to successful an
efficient operation. This provision specifies that the costs of constructin
and operating the system are to be paid for by those members whos
activities make it necessary and by those who benefit from it. Compara
tively elaborate procedures for fulfilling this directive have been deve
oped in regard to the costs of land drainage, waste disposal and wate
supply. In the course of time these procedures have come to be accepte
not only as rational and equitable but have played an important role i
the efficient operation of the system.

The area of the Genossenschaften contains notable cities such a
Essen, Bochum, Muelheim, Dortmund, Duisburg and Gelsenkirchen an
is one of the most concentrated industrial areas in the entire world—con

ining some 40 percent of German industrial capacity. The industrial omplex consists heavily of coal mining, iron and steel fabrication, and eavy chemicals. By far the dominant industries are coal, coke, iron and eel. Between 80 and 90 percent of total German production in these industries is in the Ruhr area. There are some eight million residents in the egion, which contains about 4,300 square miles of land area—roughly ne-half the size of the Potomac River watershed in the U.S.

Water resources are extremely limited if one excludes the Rhine iver (to which the Ruhr area streams are tributaries). The Rhine has a ean flow roughly comparable to that of the Ohio but is drawn upon to apply water to the Ruhr area only during periods of extreme low flow. 'here are two reasons for this. First, the Rhine itself is of very poor uality where the Ruhr enters it, and secondly, the water from the Rhine ust be *lifted* into the industrial area. With present installations, it is ossible to "back-pump" the Ruhr as far up as Essen by means of pump ations installed in dams creating a series of shallow reservoirs in the uhr. Back pumping was carried on during the extreme drought of 1959. large new reservoir for the augmentation of low flows which is nearing ompletion on a tributary of the Ruhr will even further reduce the already modest dependence of the area on the use of Rhine water.

The Ruhr area has been dependent upon the waste-carriage capacity f the Rhine to a much more far-reaching extent than for water supply. large proportion of the wastes discharged from the industrial region ito the Rhine receives comparatively little treatment. However, the contruction of a large new biological treatment plant on the Emscher (a ighly specialized stream described in more detail subsequently), will ean that virtually all effluents reaching the Rhine are given far-reaching reatment and the contribution of this area to the pollution of the Rhine vill be comparatively modest.[2]

Five small rivers constitute the water supply and water-borne waste arriage and assimilative capacity of the industrial area proper. In decending order of size, these are the Ruhr, Lippe, Wupper, Emscher and he Niers. The *annual* average low flow of all these rivers combined is ess than one third of the *low flow of record* on the Delaware River near 'renton, New Jersey, or one-half of the *low flow of record* on the Potonac River near Washington, D.C.

The three rivers serving the main industrial area—the Ruhr, the ippe and the Emscher—run roughly parallel. The Ruhr and the Wupper

---

[2]Except for saline pollution from the coal mines in the area. The Lippe carries onsiderable natural salinity and additional saline water is pumped up from the 1ines. Another major source of salinity in the Rhine is the potash industry, particu- 1rly in France. Effective arrangements for reducing salinity have not yet been made.

Rivers are mountain streams suitable for damsites and both have som
developed storage with more under development and/or planned. An i
dication of the amazing load which these rivers carry is the fact that
the Ruhr, which is heavily used for household and industrial water su
ply and which in general serves these uses well, at annual low water flo
the volume of river flow is only about 0.8 as large as the volume of wast
discharged into the river. A frequently used rule of thumb is that in ord
for a river to be generally suitable for reuse, each unit of waste dischar
must be diluted by at least eight parts of river water.

How was such an extensive industrial complex successfully bui
upon a comparatively minute water supply base with considerable atte
tion to the recreational and aesthetic amenity of water resources and
relatively modest cost?[3] Broadly the answer lies in the design and oper
tion *of a generally efficient system,* which because of the regional pu
view of the Genossenschaften, and the dense development of the are
can make far-reaching use of collective abatement measures and strea
specialization. Moreover, so-called indirect measures (waste reclamatio
process engineering and influence on the selection of industrial sites) pla
a large role in controlling the generation of industrial wastes and tl
expense of dealing with them.

In the present discussion references to the Genossenschaften may l
taken to mean either the Ruhrverband-Ruhrtalsperrenverein (RV–RTV)
the Emscher-Genossenschaft-Lippeverband (EG–LV) unless a specific o
ganization is indicated. While the EG–LV and the RV–RTV are nom
nally four organizations, the two linked pairs are each under a sing
management. These are by far the largest Genossenschaften with bot
the most complex physical water resource systems and the most sophi
ticated methods of assessing costs.

The regional system of waste disposal which these organizatioi
have established is very interesting. However, time permits only a fe
general comments to be made about it here. While the process of desig
was done with little or no explicit attention for formal optimizing proc
dures, the systems are designed and operated with the explicit objectiv
of minimizing the costs of attaining certain standards in the rivers. Mor
over, there is explicit recognition of the equimarginal principle in tl

---

[3]Despite rather impressive attention to amenities and recreation, the combine
expenditure of the Genossenschaften (which build and operate all water treatme:
plants, dams, pump stations, etc.) amounts to about sixty million dollars a year (e
clusive of capital investments), somewhat over half of which is for land drainage. T
largest water works in the area (which is a profit-making enterprise and which co:
tributes heavily to the costs of the facilities on the Ruhr) delivers water for hous
hold and industrial use at thirty cents per thousand gallons (official exchange rat
used in making conversions).

lanning procedures. Perhaps most important, the Genossenschaften pro- ide an institution which permits a wide range of relevant alternatives to e examined systematically within a functionally meaningful planning rea. However, it is also recognized that in some instances efficient solu- ons require a super-regional view. This is illustrated by the recent Ger- nan law forbidding the sale of "hard" detergents after October 1964. Sci- ntific investigation and cost assessment of alternatives leading to this egislation were primarily carried out in the laboratories of the Genos- enschaften.

In the Ruhr itself the objective of the system generally is to main- nin water quality suitable for recreation[4] and municipal-industrial water upply. In the Lippe the objective is much the same. The Emscher, by ur the smallest of the three streams, is used exclusively for waste dilu- on, degradation, and transport.

The Emscher has thus been converted to a single-purpose stream, nd is sometimes referred to as the *cloaca maxima* of the Ruhr area. t is fully lined with concrete, and the only quality objective with re- pect to it is the avoidance of aesthetic nuisance. This is sufficiently ccomplished by primary or, as the Germans say, mechanical treat- nent of effluents which largely removes materials in suspension. Since ne Emscher cannot be used for purposes other than effluent discharge,[5] ne area is dependent upon adjoining watersheds for water supply and aterbased recreation opportunities. The feasibility of this system is nhanced by the small size of the area and the fact that the streams re parallel and close together. Actually, as an aid to protecting the qual- y of the Ruhr, some of the wastes generated in the Ruhr basin are umped over into the watershed of the Emscher. By the use of plantings, entle curves of the canalized stream, attractive design of bridges, etc.,

[4]However, the quality of the Ruhr varies considerably along the course of its ow. At the head of the Hengsteysee (a shallow reservoir in the Ruhr built essentially s an instream treatment plant) the quality of the water is very poor. Neutralization, recipitation and oxidation occur in the Hengsteysee, and further stabilization takes lace in the Harkortsee, a similar instream oxidation lake. By the time the water eaches the Baldeneysee (a third such lake) the quality has improved to such an ex- ent that the water is suitable for general recreational use. This is true despite the fact nat there are further heavy discharges of treated wastes between the Hengsteyee and ne Baldeneysee. These waste discharges are generally given far-reaching treatment, equently by means of treatment plants with double biological stages (activated udge and trickling filters).

[5]One reason why such stream specialization may be advantageous is that the ate at which oxygen passes into the stream through the air-water interface is directly roportional to the size of the oxygen deficit (*i.e.* the amount by which actual dis- olved oxygen (D.O.) falls below saturation level). Thus a stream in which dissolved xygen is heavily drawn on has a much larger capacity to degrade organic wastes nan a stream with sufficient dissolved oxygen to support fish life or provide drinking ater.

care is taken to give the Emscher a pleasing appearance and to blend
gracefully into the surrounding countryside.

Near its mouth the entire dry weather flow of the Emscher up to
about 1,000 cfs is given primary treatment thus realizing scale economie
in treatment to a very far-reaching extent. The heavy burden put on the
Rhine both from upstream sources and from the Ruhr industrial are
(largely via the Emscher) has caused great downstream costs.[6] Conse
quently the Emschergenossenschaft is now laying detailed plans for bio
logical treatment of the Emscher. An experimental plant indicates the
this treatment will be highly successful. When the new plant is built, the
contribution to Rhine river pollution on the part of the Ruhr industrie
area will be substantially mitigated and the area will come quite near to
being a closed water supply-waste water system.

While formal optimization procedures have not been utilized by
the Genossenschaften, they probably have realized the major gains from
viewing the waste disposal-water supply problem as one of a system char
acter rather than solely as a matter of treating wastes at individual out
falls. They have made extensive use of scale economies in treatment by
linking several towns and cities to a single treatment plant when the
costs of transporting effluents to the plants were less than the saving due
to additional scale economies that could be realized—in the case of the
Emscher they have, in fact, linked an entire watershed to a single treat
ment plant. They have utilized opportunities for more effective treatmen
of wastes which accrue through combining industrial and househol
wastes. They have made use of stream specialization for recreational an
water supply purposes, and artificial ground water recharge for qualit
improvement purposes. They have at various times and places used flov
augmentation, and direct aeration of streams. They have explicitly cor
sidered the differential ability of streams to degrade wastes at variou
locations (resulting from the opposing effects on oxygen balance of wast
degradation and natural reaeration) both in determining location of trea
ment plants and in influencing the selection of industrial plant site:
Where scale economies, or special technical competence of Genosser
schaft staff,[7] merited it, they have established their own waste recover

---

[6]Especially in Holland where even recently introduced large-scale ground wate
recharge projects are failing to supply suitable quality water.

[7]Administrators of the Genossenschaften place great emphasis on the economie
which result from a single staff planning, building, operating, and supervising th
water resources facilities of an entire basin. The Ruhrverband operates eighty-fou
effluent treatment plants (to which, on the average, four new ones are added eac
year), four large detention lakes, twenty-seven pumpworks, 300 km of trunk-sewe
run-of-the-river power plants, six dams (one in addition is under construction), powe
plants associated with the dams, and their own electricity distribution systems with
total staff (including laborers, apprentices and janitorial help) of 780 persons.

lants (phenols). They have, in other instances, induced waste recovery
r process changes by levying charges for effluent discharge based on
quantity and quality of waste water and by acting as a cooperative mar-
eting agency for recovered waste products. While not always in a comprehensive manner, decisions between different alternative ways of
chieving objectives have, at least in a rough and ready fashion, been
based upon consideration of cost "tradeoffs" between them. Finally, and
of considerable importance, they have provided for monitoring of the
treams (especially those used for water supply) and operation of facili-
ies to take account of changing conditions in a more or less continuous
ashion.

Whether, in view of the value of recreation use, the conversion of the
Emscher into an open sewer, or the very heavy use of waste degradation
capacity on certain stretches of the Ruhr is optimal, cannot be determined because explicit valuations of recreation use have not been made.
Actually, outdoor recreation is quite impressively catered to in the Ruhr
area. This is probably due to the considerable power which the communities and counties exercise in both the water Genossenschaften and the
Siedlungsverband (the latter is the agency responsible for land use planning in the Ruhr area and eighteen major cities are in its planning jurisdiction). Coordination of the work of the Siedlungsverband and the
Genossenschaften is significant not only in producing an explicit weigh-
ng of recreational and aesthetic values against others in the development
and use of water resources, but also in providing explicit consideration of
ndustrial location (especially with respect to areas of compact industrial
development) as a variable in water use and waste disposal planning. In
other words, planning procedures which provide for cooperation between
the staffs of the water and land use authorities make sure that a variety
of costs and benefits involved in patterns of land use and water quality
alteration are reflected in the planning process, even though neither
takes a comprehensive view in all its details. Moreover, once a general
pattern of development is laid out, the water authorities continue to influence more specific industrial location decisions as well as process engineering and waste recovery through their systems of charges.

PART III—COST ASSESSMENT PROCEDURES

It is of course clear that solely from the point of view of resource
allocation, the method of pricing adopted, or indeed whether effluent
charges are made at all, would be relatively unimportant if waste loads
delivered to the system were unresponsive to the charges imposed on

them. This would appear in general to be the case with respect to house
hold effluents, since there is comparatively little households can do t
diminish their waste loads. However, through product and process ac
justments, through waste recovery, through separation of wastes an
various forms of pretreatment, industrial waste loads can be altered ove
very wide ranges.

Numerous adjustments, especially in process design, are being mad
by the industrial plants in the Ruhr area as a consequence of the Genos
senschaften's methods which force industry to bear at least a significar
portion of the social costs of waste disposal. In general indirect method
of reducing wastes such as recovery and process changes are considere
on a par with treatment in the work of the Genossenschaften. This is on
of the reasons why the giant North-Rhine-Westphalian industrial com
plex can operate on a water base which is tiny by United States stand
ards.

The vast variety and potentially large effect which recovery proc
esses and process design can have on waste loads point to the importan
role which a correctly planned system for the assessment and distributio
of costs may play in restricting to an optimum degree the amount o
waste products which industrial society produces. This being the case,
closer look at the methods of cost assessment used by the Genossen
schaften is in order.

PART IV—COST ASSESSMENT AND EFFLUENT CHARGES

A. Description

The question confronting both these organizations is how the di
verse wastes produced by industrial enterprises can be assessed with a
appropriate portion of costs.

Very briefly put, the Emschergenossenschaft procedure is roughl
as follows:

1. There is estimated first an amount of water necessary to dilute a
given amount of waste materials subject to sedimentation (no distinctio
is made between organic and inorganic material) in order that they no
be destructive to fish life under the conditions of the area. An amount o
dilution water required by such materials in a given effluent is then com
puted on that basis.

2. An analogous calculation made for materials subject to biochem

al degradation (and which therefore exert an oxygen demand) but which are not subject to sedimentation.

3. The amount of dilution required under specified conditions in rder that the toxic material in the effluent not kill fish is computed by irect experimentation.

4. Certain side calculations having to do with water depletion, heat 1 effluent, etc., are made. The derived dilution requirements are added ogether for the effluent and form a basis for comparison with all other ffluents. In principle, costs are distributed in accordance with the pro-ortion of aggregate dilution requirements accounted for by the specific ffluent. One might say that this procedure is based on a particular phys-cal objective, *i.e.* not to "kill fish." However, the result of the method is sed as an "index" of pollution even when effluents are discharged to treams in which lower or higher standards prevail than needed to pre-erve fish life.

The Ruhrverband method is also based on a physical objective but n a different one. Again, the details are described elsewhere. In essence, owever, the method is founded on the concept that toxic wastes by kill-ng bacteria and slowing down the rate at which wastes are degraded ave somewhat the same effect on treatment plant effluents and the level f BOD in streams as an *increase* in the amount of degradable material. On the basis of laboratory tests, an equivalence is formed so that toxic as vell as degradable wastes are converted into a standard unit—a "popula-ion equivalent BOD."

These procedures can be criticized on technical as well as economic grounds. They are indeed recognized as less than ideal by the Genossen-schaften but are generally defended as being readily understandable and elatively inexpensive to administer. Further development of cost assess-nent procedures along even more meaningful lines appears possible, lowever, especially in light of new technology which has recently be-come available or is now under development and which points to an easing of certain measurement problems.

Three areas will be briefly commented upon:

1. A formula for the assessment of costs based upon a physical objective will tend to lead to some misallocation of costs.

2. The procedure distributes average costs rather than assessing marginal costs.

3. Important economies might be achieved by an application of peak load pricing.

B. Comments

1. Deficiencies in the use of physical objectives in cost distribution
The minimization of costs either in the limited sense of minimizing cost
for given objectives or in the broader sense of minimizing the social cost
associated with waste disposal, has been taken to be the general objective
of a regional water quality control system. This objective logically implies
that both the system design criteria and the cost distribution criteria must
be based on *costs*, not directly upon *physical effects*. Thus in the matter
of system design it is *costs* which must be balanced at the margin.

Similarly, in the matter of cost distribution, a procedure based upon
physical results alone (as are the Emschergenossenschaft and the Ruhr
verband procedures) will, in principle, not allocate costs properly. For
example, a substance may be very destructive to fish (and thus merit a
high weight in the Emschergenossenschaft's method) but be relatively
inexpensive to treat or to deal with by other means. In this instance a
disproportion between the costs of dealing with waste substances and
costs allocated to those producing them will arise. This would be true
even if the sole objective of the system were to avoid killing fish.

In the broader context of multipurpose water use a practical illustra
tion of substances which are not handled at all adequately by either
method is the phenols. Phenols are produced in the distillation of petro
leum and coal products. In low concentrations (a few parts per million)
they do not exert much oxygen demand nor are they very toxic to fish
But even in the smallest concentrations they pose the most serious prob
lems in the preparation of drinking water. When water containing minute
amounts of phenols is chlorinated in order to kill bacteria, extremely evil
tasting chlorophenols are formed.

Phenolic substances present a problem in all the world's great in
dustrial complexes. The recommended limit for phenols in the latest revi-
sions of the United States Public Health Service recommended drinking
water standards is an infinitesimal 0.001 mg/l. Unless water supplies
drawn from surface sources are to be very unpalatable, phenols must be
kept out of waste water or else removed at great expense by the application of activated carbon at the water plant. Moreover, very small amounts
of phenols in streams can impart an unpleasant "carbolic" taste to fish
which may as effectively destroy their value as though they were actually
killed. If a stream standard is imposed (say, instead of attempting to balance the cost of damages and abatement at the margin), this may mean
far-reaching treatment or recovery processes are necessary for the phenol

containing effluents. Clearly the cost of those procedures and/or damages is not reflected in the allocation procedures used by the Genossen-chaften. This follows because in one instance the implied objective of the cost allocation procedure is not to kill fish and in the other not to kill bacteria.

Where substances do involve a large disproportion between the costs they impose on the system and the costs allocated to them by the methods, side calculations are generally made to take this aspect into account. The basic point, however, is that the cost assessment procedure is not fully consistent with the design objective.

2. *The appropriate concept of costs.* As earlier indicated, in prin-ciple, charges made for effluent discharge should equal full opportunity costs including increased water supply costs and foregone productive op-portunities downstream. If there is no separate system design problem, imposition of these costs on waste discharges would tend to produce an allocation of resources which maximizes social product.

If scale economies introduce a separate problem of system design, levying only the (appropriately defined) costs of a correctly designed sys-tem would tend to accomplish the resource allocation objective. The ap-propriate process, product and location adjustments would be induced even though the costs of residual damages (which most probably would exist in an optimum system) are not specifically imposed on the waste discharger. This remains true even if the objective is to accomplish min-imum cost of achieving certain standards. Consequently, basing the charges on costs actually incurred by the system as the Genossenschaften do, is not *per se* incorrect.

A number of different possibilities exist with respect to the elements of a system which are directly constructed and operated by a regional authority and those left in the hands of fiscally independent decision makers. However, in general a charge set equal to marginal cost, at their optimum level of utilization, of those abatement measures directly un-dertaken by the authority will be appropriate.[8] (In essence each dis-charger is subjected to a "last added" test). This is because the marginal costs of induced measures will tend to be equated with the charge per unit of waste discharge and consequently with the marginal cost of other alternatives.

It is clear, however, that the Genossenschaften distribute average costs to the dischargers, not marginal costs. Thus, if the system displays declining costs the charges as levied will generally be too high, whereas

---

[8]If some action is taken by the authority for each pollutant.

if increasing costs are incurred, the opposite will generally be the case
The efficiency argument against average cost pricing in this context i
that it will not induce appropriate use of measures controlled by fiscall
independent decision makers.

3. Cost variation over time.   For purposes of planning, costs prop
erly include all capital costs of new facilities. The system of abatemen
(ideally) should be expanded to the point where an additional unit o
abatement (optimal combination of measures) for any given pollutant, o
combination of pollutants, raises the expected present value of tota
abatement costs as much as it diminishes the expected present value o
pollution damages. Or if there is a constraint, or a standard, until th
constraint is met and the expected present value of marginal total cost o
all alternative measures for achieving the constraint are equalized.

However, at any given time with an established abatement system
only current operating costs and opportunity costs internal to a multipur
pose system are relevant. These systems costs tend to vary strongly ove
time due to hydrological variability. This is true whether the objective i
to equate marginal abatement costs and marginal damage costs or t
minimize the costs of meeting a standard. This suggests the possible de
sirability of attempting to base prices on a rather short-run variation i
costs.

The cost distribution methods of the Genossenschaften fail to tak
account of the fact that waste disposal costs (in the broad sense) ar
highly variable through time. As already mentioned, this variability re
sults from the fact that the dilution and degradation capacities of stream
show strong variation over time. For example, the long-term average dis
charge of the Delaware River at Trenton, New Jersey, is about 12,00(
cfs, the mean annual flow is roughly one-fourth that and the low flow o
record about one-tenth of the average flow. The average flow of th
Ruhr is about 2,600 cfs and the average annual low water flow is onl
about 140 cfs. If the costs of waste discharge to the polluter do not var
in accordance with flow, there is no incentive to reduce discharges durin
low flow periods, even though the concentration of pollutants, attendan
damages and the costs of optimally operating abatement works tend t
rise sharply during low stream stages.[9]

[9]Oxygen conditions are especially likely to deteriorate radically during suc
periods because high temperatures (which cause bacteriological activity to increas
and the oxygen saturation level of water to decline) ordinarily correspond with lo
flows. The combination of high concentrations of toxins and low oxygen levels ca
easily be fatal to fish. If oxygen levels are below several parts per million, oxygen de
ficiency itself will kill fish and if oxygen becomes exhausted extreme nuisance condi
tions accompany the development of anaerobic processes in a stream.

If it were clearly uneconomical to change the amount and/or qualty of waste discharge over short periods of time, it might not be a matter f great concern whether or not costs levied upon polluters varied corespondingly over time. However, it appears probable that measures to hange the pattern of discharge would enter economically into a quality ontrol system designed to minimize costs. For example, depending upon he location of a manufacturing concern and upon attendant land values, t may be less expensive for the company to withhold its waste discharge emporarily in a lagoon rather than bear its share of the costs of storing nd, at long intervals, releasing a much larger volume of river water. In ome instances, especially where the product is storable, it may be more conomical to reduce or halt production during low flow periods rather han to provide additional treatment or dilution capacity for an unhanged effluent. In other instances it may pay the manufacturer to proride temporary treatment (like chemical neutralization of acids) rather han meet the full costs of putting his effluents into the receiving water luring low flow periods. In the light of such possibilities, incentives hould be provided to use them to an optimal degree.

The method of peak pricing has been extensively carried out by ·lectrical utilities, hotels, resorts and theaters, among others. It regards he service performed (in this case pollution abatement) as different ac·ording to the heaviness of the load on the system. In the case of elec-

---

The concentrations of substances which alter Ph value of water, affect its hardιess, create tastes and odors, cause dissolved solids content to rise—all tend to be ιigher during periods of low flow. This leads to rising municipal and industrial water upply treatment costs and a variety of pollution-caused damages to facilities and ·quipment. The point is that the social costs of pollution rise strongly during periods ɔf low flow.

An authority controlling a going waste disposal system and attempting to operιte it in such a way as to equate the relevant marginal costs would necessarily incur nuch greater costs of operation during low flow periods than during higher flows.

During the critical periods such measures as increased aeration in activated]ludge plants and the addition of chemicals which aid precipitation in all types of ·reatment plants might be undertaken, leading to an increase in operating costs. Durng such periods artificial reaeration of streams may be done directly in the stream or ·hrough the turbines of hydro plants which in the first instance involves direct costs ιnd in the second, indirect costs because power plant efficiency is cut down. During ow stream stages the augmentation of flow from reservoir storage offers the opportuιity to increase waste dilution and if there are alternative uses for the stored water ('say, peak power generators, or for recreation), an opportunity cost (internal to the water resources system) is incurred.

It is thus quite clear that costs are strongly related to time of discharge. In fact ·he social costs of a given quantity of a pollutant discharged at one time may easily ɔe a multiple of those at another. Indeed, during periods of high stream flow waste]disposal into the stream is likely to be virtually without downstream damages. During ɔuch periods the only justification for operating treatment plants at all may be to avoid the aesthetic nuisance of floating materials in the water.

trical utilities, this necessitates a price varying by time of day and the season. In the case of pollution probably no more than seasonal variations could be justified.

In assessing the merits of applying "peak load" pricing principles to effluent discharges, the costs of determining variation in the quality and quantity of effluent over the relevant period must of course be considered. Presently, the Genossenschaften generally establish by sampling an effluent quality which is taken to be typical for the year. The quantity effluent discharged during the year is ordinarily based upon measurements reported by the plant.

The costs of operating analysis sampling programs along present lines would mount sharply if an effort were made to determine quality and quantity variation with sufficient continuity to permit peak load pricing. Fortunately in recent years progress has been made in the development of automatic monitoring devices. Such variables of river quality as dissolved oxygen, acidity-alkalinity, salinity, specific conductance (dissolved solids), temperature and turbidity can be continuously measured with fairly simple devices. Some measurements are already successfully carried out on effluents and there is considerable promise that others can be developed. Optimism appears justified that accurate and comparatively simple devices for continuous measurement of a wide variety of water quality characteristics can be worked out.

In some instances an occasional rather thorough laboratory test may have to be done in order to establish relations which will permit the use of certain continuously measurable characteristics as surrogates for those which can only be measured with difficulty. For example, if fish toxicity is an important variable (as it is for example in the Emschergenossenschaft cost allocation method), a relation may be established between certain measurable substances in the effluent and fish kills; thus over specified periods of time this would permit the surrogate measurement to substitute for a direct toxicity test.

It is probably not excessively visionary to foresee a time when a number of important quality characteristics can be continuously recorded, at a central point, for every major outfall in an entire basin at comparatively modest cost.

A river dies — and is born again

FRED J. COOK

The author is a free-lance writer who lives in New Jersey, near the Raritan.

The Raritan River at New Brunswick, N.J., was a gay patchwork of color. People crowded the banks and boats bedecked with bunting lined the course where some of the world's leading swimmers were competing in New Brunswick's annual swimming festival. It was the summer of 1922. As the spectators cheered and boat whistles sounded, Gertrude Ederle, famed as the first woman to swim the English Channel, churned home to a new world record for the 440-yard free style against Britain's best, Hilda James.

Old pictures capture something of the gaiety and triumph of that moment—the sun bright, the boats picturesque, the river beautiful to look at. But the competing swimmers were pulling wry faces as they came out of the water. "It tastes like acid," some of them said.

They were right. The Raritan, which John Davis, an English poet, had described in 1806 as the "Queen of Rivers," was being transformed into the Queen of Sewers. The summer swimming festival of 1922 was the last that could be held in its waters.

This is the story of the life and death—and partial resurrection—of a river. It is a case history of national significance, for water pollution today is one of the most acute of our domestic problems. According to the Department of Health, Education and Welfare, of the 11,420 communities that have sewer systems, 2,139 (18 percent) still discharge untreated

From the *New York Times Magazine*, April 18, 1965, p. 22 © 1965 by The New York Times Company.

wastes into the nation's waterways. Federal action in a drive to clean them up has been taken in 600 cities and a like number of industries, affecting 7,000 miles of rivers, lakes and bays.

The struggle to preserve and restore the Raritan has been, in a way, a national pilot project and the techniques used have been widely copied. A colored motion picture, produced in 1951 to carry the message of water pollution to the people of the Raritan Valley, is still one of the principal weapons in the arsenal of the Health, Education and Welfare Department. It was used again recently, and used widely, in the District of Columbia as a part of the campaign to clean up the Potomac River.

The tragedy of the death of a river can be appreciated only by contrast with its former glories. The Raritan rises in the mountainous northwest of New Jersey, one of its tributaries stretching to within six miles of the Delaware. It winds and loops for more than 100 miles in a generally southeasterly direction, nearly bisecting the state, until its mile-wide estuary debouches into Raritan Bay between Perth Amboy and South Amboy. It is a placid river, flowing between wide meadows, low rounded hills or high bluffs of reddish shale.

An early Dutch settler described the Raritan Valley in 1640 as "the handsomest and pleasantest country that man can behold." A little later, an English settler raved about the abundance of fish and wild fowl and succulent "oysters I think would serve all England." During the Revolution, George Washington, hard-pressed by the British, is said to have reined in his horse upon a bluff above the river and exclaimed: "This is the loveliest scene on earth! If rest and repose ever be my lot, I should, next to Mount Vernon, choose this peaceful spot in which to pass my life."

It was the same even in the early nineteen-hundreds. In the spring, great schools of shad migrated up the Raritan to spawn, and fishermen stretched their nets halfway across the stream. Raritan shad were so plentiful they sold for a quarter apiece and were packed and shipped to distant markets. Shellfishing in the river and bay was a million-dollar industry; Amboy oysters were considered a delicacy. Sailing ship and barge captains used to bring their vessels upriver to fill their water casks because, they said, Raritan water tasted better and kept pure longer than any other.

Then the river died—or rather, was killed almost overnight. In the years after World War I, industry came to the Raritan Valley and towns grew into cities. In 1901, only 6,700 people had been employed there in manufacturing plants; by 1927, the number was 17,500. And the total has kept on growing. By 1952 there were 360 manufacturing establishments in the valley, compared with forty-eight in 1901. The wastes from breweries and firms producing copper, lead, paint, asbestos and chemicals, combined with the sewage of an increasing population, turned the Raritan into an open sewer carrying stench and contamination to the sea.

What happens when a river dies? In this case, it became so clogged with wastes that their decomposition absorbed virtually all the oxygen in its water upon which aquatic life depends. Yeast wastes, sulphur wastes, phenols, acids, alkalis were all pouring into the Raritan and its tributaries. Some of the deadly acids killed fish outright, but by far the greater damage was done by the chemical deoxidation. Algae died, bait vanished, the shellfishing industry was wiped out. Ducks in great flocks no longer came, for their natural foods had been destroyed; there was nothing for them to eat.

By September 1928, *The Home News* in New Brunswick could state that "the Raritan River is known today to be the filthiest stream in our whole country." Health standards, it was pointed out, permitted only twenty-five colon bacteria per cubic centimeter of water, but readings taken at New Brunswick had gone as high as 1,800 and in other places to 2,200. It was much too late to save the river, and efforts would have to be made to revive it.

The first attempt—one that showed how *not* to meet the problem of river pollution—was piecemeal. An engineering report gave the municipalities in the Raritan Valley a choice: they could build their own individual treatment plants for their waste products or they could join in a cooperative effort to construct a trunk sewer line and a central treatment plant to handle the wastes of the entire valley.

Local pride and prejudice being what it is, the first method was preferred. Federal aid was obtained from Roosevelt's Public Works Administration and by January 1936, municipally owned sewerage plants up and down the valley were under construction. But hardly were they finished before it was obvious they could never solve the problem. In 1940, the New Jersey Department of Health was citing New Brunswick for the inadequacy of its new treatment plant.

World War II not only halted the construction of plants but also greatly stimulated industrial growth and thus the movement of population into the valley. The overburdened municipal plants simply could not cope with this rising flood of wastes and pollution in the Raritan became as bad as it had ever been. South River, for instance, one of the main tributaries of the Raritan, was a veritable sinkhole. H. Mat Adams, former New Jersey conservation commissioner and now chairman of the Middlesex County Sewerage Authority, recalls: "The pollution of the South River and the fumes rising from it were so bad that paint peeled off houses anywhere near its banks, and even wallpaper and paint inside the houses were stained."

Elaborate engineering studies were again made, and in August,

1950, Middlesex County created a sewerage authority to attempt to carry out the recommendations. This time they called unequivocally for a trunk sewer line and a central treatment plant. Such a system was not only more efficient technically but could be designed to take care of the needs of the area until 1975, at which time it could be readily expanded to handle the problem until the year 2000.

The basis of the trunk sewer system was to be a massive sewer line running the length of the valley, with branch lines bringing in the wastes from municipalities and industries all along the Raritan and its tributaries. Funneled into this main line, all the sewage would be pumped to a central plant in Sayreville for processing. Here it would be treated mechanically and chemically before being pumped two miles to an outfall in Raritan Bay. Experts from the Woods Hole Oceanographic Institute had selected the outfall site, which was precisely where the sweep of tides would carry the discharge straight out to sea. Solids not broken down in the treatment process would be taken away by barge and dumped twelve miles out in the Atlantic. The engineers predicted that this all-inclusive system would successfully purify the Raritan and its tributaries.

Such was the vision. To get everyone to accept it was the real, the monumental problem. The law, the power of state and county authorities, was no help. The State Health Department could set health standards and seek injunctions against industrial plants or municipalities polluting the Raritan, but there was no authority to compel everyone to join in one cooperative, valley-wide clean-up. Persuasion was the only weapon.

But anyone who has seen borough councils in action knows how jealous they are of their own prerogatives, how petty the issues that often divide them. And in the Raritan area, there was the additional stumbling block that the valley's towns and cities had spent millions of dollars building their own treatment plants only ten or fifteen years before. Was all that investment to go for nothing?

Fortunately, the businessmen of the valley had become committed to the cause. Robert Wood Johnson, then head of the pharmaceutical house of Johnson & Johnson and one of the prime movers in the campaign, was joined by other prominent industrialists and together they produced the color film mentioned earlier, stressing the dangers of pollution and explaining the proposed solution. The Raritan Valley Junior Chamber of Commerce supplied projectionists and speakers, and showed the film at more than 1,200 meetings. It proved most effective.

The movie opens on a newspaper headline announcing the State Department of Health's ruling that no new industrial plants can be built along the Raritan with the idea of using its water. It shows a child, sick in bed with a disease contracted from polluted water. It tells the story of the "Sad Shad" who used to come to the Raritan to spawn but who now, a clothespin on his nose, shuns its filthy waters. That filth is vividly depicted. The camera focuses on open pipes spewing out untreated sewage; it shows the turgid tide clotted with great blobs of sludge like dirty ice floes—and it shows young boys, stripped to the waist, answering the irresistible call of running water in the springtime, plunging from a tree-shaded bank into the Queen of Sewers.

Newspapers, radio and television stations joined in the publicity campaign. The cartoon figure of the Sad Shad was used for further disclosures about the Raritan's pollution. But although these public relations techniques helped make the public aware of the problem, the obstacles of local pride and narrow self-interest still had to be surmounted. Mat Adams realized that if the Sewerage Authority's trunk sewer plant was to be adopted, some way would have to be found to compensate the municipalities for their large but ineffective investments in individual treatment plants.

Adams consulted some of the privately owned and operated water and sewer companies and a formula was worked out. The Middlesex Authority agreed to remunerate every municipality that joined the trunk sewer plan for the sewerage plant it would no longer need. After calculating depreciation, the authority would pay the town the estimated value of the plant, which would, however, remain the property of the municipality. The cost of this was estimated at close to $1 million, but Adams and the other commissioners realized that otherwise the authority would have no chance.

There were other persuasive arguments. As the authority would be privately financed through the sale of bonds, there would be no new burden on the taxpayers. In contrast, if the towns of the valley were to modernize their own overburdened plants, some $43 million would be added to the local tax bill. The cost of operating the trunk sewer line would also be less—$1.4 million compared with $2.1 million a year for the municipal plants—a clear $700,000 saving. Furthermore, the total cost of using the trunk line and central treatment plant would be apportioned, with municipalities paying only about 35 percent to industry's 65 percent.

With the problem so acute and the solution so well programed, the trunk sewer proposal won wide, though not complete, acceptance. Some industries and municipalities, having invested millions in their own treatment plants, simply refused to go along and there was no way to compel them. At the mouth of the river, Perth Amboy and South Amboy stayed out, and along its upper reaches, municipalities like Somerville and Manville and companies like Johns-Manville and American Cyanamid Co. withdrew from the plan. The rest of the valley signed up, but even then, acceptance was less than complete in some instances. A number of plants insisted on treating portions of their own waste and the sewerage authority drew up contracts permitting them to do so provided the effluent met state health standards. With these exceptions—and they were to prove serious—the trunk sewer system went into operation on Jan. 14, 1958.

The results were striking. The system was only about six months old when Sol Seid, the authority's chief engineer, began getting telephone calls from residents along South River.

"What are you people doing to the river?" the callers demanded angrily. "It's all covered with great black globs of stuff."

Seid and Mat Adams chuckle at the recollection. "We sent our engineers out right away to check on what was happening," Seid says, "and we found that the great jet-black 'globs' covering the river were sludge that had broken loose from the bottom. We had estimated that it would take about a year for this to happen, but already the river was beginning to cleanse and purify itself." What had happened was that the stream, in its movement was no longer adding to the deposits of sludge; instead, it was aerating them and the rush of purer water was breaking them loose and carrying them away.

By September 1958, the local press was reporting that crabs had returned to the South River. "The crabbing craze has hit South River, Sayreville and nearby communities with the impact of a small gold rush," one story declared. Fish were seen for the first time in years in the main channel of the Raritan. There were snappers along the New Brunswick waterfront and a school of striped bass was spotted off the municipal docks. Boatmen, who only the year before had complained that their propellers churned up water black as mud in mid-channel, now admired their clear white wakes. But the battle to restore the Raritan had been only partly won.

Today, nearly seven years later, the battle goes on. Since the Raritan drains an area of 1,100 square miles, towns in Somerset and Union Counties have joined the trunk sewer system and a representative from each of those counties now sits on the Middlesex authority's nonsalaried board. At the same time, the growth of industry and population in the valley has outstripped all estimates. It had been predicted, for instance, that Edison would have a population of 16,000 by 1960; in fact, it had 44,000. Madison, with 10,417 people in 1950, grew to 15,122, and other municipalities showed the same phenomenal expansion. A severe strain was thus placed on the trunk sewer system, but the authority anticipated the trouble and

ast year increased the capacity of its central treatment plant from fifty-
.wo to seventy-eight million gallons a day. But the Raritan is still not the
pure stream everyone had hoped for—though it has tremendously improved since the days when it earned its title of the Queen of Sewers.

"There is some swimming in the South River and the Raritan, but
we don't encourage it," Mat Adams says. "The water is still not pure
enough."

Why not? In the view of Adams and other pollution experts, a 100-
mile-long river cannot be cleaned up on a hit-and-miss basis. Although
constant testing by the sewerage authority and the U.S. Public Health
Service has shown that the effluent from the central treatment plant con-
sistently betters the minimum requirements of the State Department of
Health, the Raritan remains considerably polluted.

Adams says bluntly that the conception of a completely purified
Raritan "went out the window" when the idea of a true riverwide authority was defeated and certain municipalities and industries "went their
devious ways in solution of their waste-treatment problems."

Constant checking (the authority made some 50,000 laboratory analyses in 1963 alone) has shown some minor pollution still seeping into the
Raritan from industries handling portions of their own wastes, and stiff
warnings have been issued to the violators. But the more serious sources
of pollution, in Adams's view, are outside the authority's jurisdiction. The
river estuary between the Amboys, with a tidal flow extending eighteen
miles upriver, is constantly polluted by waters sweeping in from New
York harbor and Raritan Bay. And upstream, in the area of municipalities
and industrial plants that refused to join the trunk sewer system, there is
still pollution coming down that undoes much of the good work.

The authority has been pressing the State Department of Health to
set more rigid pollution standards and to see that they are enforced—
but, "We're impatient," Adams says frankly; "we're dedicated to a job
and because of things beyond our control, we can't get on with it. You
can say that we're getting mighty impatient."

Adams recognizes, of course, that the task is not simple. Outside his
authority's jurisdiction, in the areas of the most serious pollution today,
only the State Health Department can act. Its job would be greatly reduced if all waste disposal in the valley had been tied into the trunk
sewer system as originally proposed, but with the upriver municipalities
and industries going their own way, the Health Department has a serious
problem. Before it can go to court for an injunction and clean-up order
against a polluter, it must first have irrefutable, scientific proof, difficult
to obtain, to answer the defendant's inevitable argument that it is not his
fault but the other fellow's. Even then, as cases brought against polluters

along the South River in the 1940s had demonstrated, the legal entangle
ments are long and frustrating—and, all the while, the pollution continues
Mat Adams understands this.

"People instinctively seem to have a negative attitude," he says
"You have to show them how vitally pollution affects their every interes
—their health, their jobs, their recreation, their whole way of life. Only i
people are made to realize this can you get action."

"For example, they have just one engineer and one deputy assigned
to the entire Middlesex-Monmouth County area," Sol Seid explains. "Jus
two men can't do the whole job; it's impossible. To enforce health regula
tions in regard to water pollution, you have to be able to track the pollu
tion to a specific source—and you have to have the kind of evidence tha
will stand up in court. This takes manpower—scientific manpower—and i
can't be done unless the Governor and Legislature give the departmen
adequate funds to work with."

Those funds will become available, Mat Adams believes, only whei
the public becomes sufficiently aroused by the problem to demand actior
He thinks that President Johnson and Governor Rockefeller, by advocat
ing broad antipollution programs, have helped to stimulate this kind o
public interest, but more is needed.

"Public officials, I am sure, want to do this job," Adams says, "bu
they are not going to call for the kind of money the task requires unles
they get some assurance of popular support. The people are the ones wh
are going to have to demand clean water."

Pittsburgh: how one city did it

TED O. THACKREY

Ted O. Thackrey is the editor of Better Times, *published by the Community Council of Greater New York.*

It's possible in Pittsburgh, Pennsylvania, these days to wear a white shirt, or a white dress, from dawn to dusk and come home almost spotless. Housewives are able to hang out a wash on a sunny morning in Pittsburgh, if they are so minded, without having the load to do over by noon because of sootfall and dust. The downtown skyscrapers, or most of them at any rate, appear newly constructed, though it may have been five years or more since the surface was washed or blasted. Since the middle or late 1940s, a generation has grown up in Pittsburgh that on any day could look up and see, of all urban wonders, the sky.

Before the 1940s, however, generations lived in Pittsburgh—once known as the Smoky City—that often found it difficult to see the opposite side of the street. Today Pittsburgh, the once murky steeltown at the junction of the Allegheny and Monongahela Rivers, is one of America's cleaner cities. It has, for example, only half the dustfall registered in New York City: thirty tons per square mile per month on the average, compared with New York City's current sixty tons.

The story of the Pittsburgh renaissance reaches back over more than three decades and is by no means concluded. It has been and continues to be a bitter, emotionally charged battle to keep the atmosphere pure enough to breathe and clean enough to see through. The struggle to rid the mighty rivers of pollutants and to halt the despoiling of the land has barely begun.

Reprinted with permission from the *Saturday Review*, May 22, 1965, pp. 46–47.

Coal and steel interests, along with the railroads, were not only th[e] job-giving lifeblood of Pittsburgh but were also the chief despoilers o[f] air, land, and water in the old days. But Pittsburgh also had many me[n] of determination and vision who foresaw that Pittsburgh could choke it self to death on the smoke and fumes belching from factory and locomo tive chimneys. The fight for clean air really began late in 1939, but it wa[s] seven grimy years before any specific results could be seen.

Leading industrial and business executives, including representa tives of the steel industry, sparked the renaissance through formation o[f] the Allegheny Conference on Community Development. By 1942 anothe[r] civic group, the United Smoke Council, was formed by 142 men an[d] women from seventy-one organizations. World War II was on—and th[e] United Smoke Council entered the battle on patriotic grounds; smok[e] control, it proclaimed, was essential to save fuel needed for war produc tion.

By 1943 an antismoke ordinance had been passed. It provided tha[t] as of October 1 of that year private homes were to burn only smokeles[s] coal, fed to furnaces by stokers. But the effective date of the ordinanc[e] was eventually postponed to March 1945, partly because of oppositio[n] from soft-coal interests and partly because there weren't enough stoker[s] and other smoke control equipment to go around. Meanwhile, rallies[,] meetings, and membership campaigns in the fight for smoke control wer[e] almost continuous. Finally, in 1945, the United Smoke Council became [a] part of the Committee on Smoke Abatement, a subdivision of the Al legheny Conference on Community Development.

But something else took place in 1945 that dramatized the struggl[e] for clean air so prominently that most of Pittsburgh's population had t[o] take sides: David L. Lawrence, an independent and well-to-do civi[c] leader, campaigned for the Democratic nomination for mayor, and hi[s] major appeal was to those who wanted smoke control.

Not everyone wanted smoke control by any means. Householder[s] were told that smokeless fuel would cost them dearly and that smok[e] control would drive industry from Pittsburgh; that it would become [a] jobless ghost town if the Lawrence supporters had their way. "Remem ber Little Joe," read the opposition placards. ("Little Joe" was a per sonification of "the little man" who would have to buy smokeless coa[l,] stokers, and other conversion equipment if smoke control laws were en forced.) It was a tough battle, but Lawrence won the nomination and th[e] election. And he proved to be a man who kept his campaign promises[,] particularly where smoke control was concerned.

One key to Lawrence's campaign was an impressive fact sheet abou[t] the high cost of atmospheric pollution. Respected physicians said flatl[y]

hat Pittsburgh was not a healthy place to live, and statistics on respira-
ory diseases appeared to bear them out (in 1945, for example, Pittsburgh
ed the nation in deaths from pneumonia). It was disclosed that forty in-
lustrial firms had decided to leave Pittsburgh because of smoke, smog,
ınd impure air, as well as because of a recurring danger from floods and
·ontaminated water. A department store estimated that on one winter
lay alone smog soiled merchandise and home furnishings so seriously
hat the loss in cleaning bills and markdowns was $25,000. Nevertheless,
he coal industry and the railroads continued to oppose clean-air efforts
ın economic grounds.

In 1946, the Pittsburgh Chamber of Commerce, the League of
Vomen Voters, and other civic groups joined the Allegheny Conference
ight to extend air-pollution efforts to all of Allegheny County. Pittsburgh
ıad discovered what many other regions have still failed to grasp: that
ıolluted air knows no political boundaries; it drifts over borders and
:hokes communities impartially. Mayor Lawrence was deep into the fight
ıy now, meeting around the clock with industrial leaders. In 1947, during
ı stalemate in the General Assembly in Harrisburg over legislation to per-
nit countywide control of air pollution, one of the major opponents capit-
ılated. The Pennsylvania Railroad, which not only hauled but burned
:oal, feared a loss of freight tonnage and was also worried about the
·normous cost of converting coal-burning locomotives to Diesels. At that
ıoint Richard King Mellon, most powerful of the city's financial and in-
lustrial giants, was persuaded to use his influence with the Pennsy's di-
·ectors. The opposition died away, and the legislation passed.

By this time Pittsburgh was basking in some of the benefits of air
ıollution control. Two years after its smoke control ordinance went into
·ffect, visibility in downtown Pittsburgh had improved 67 percent, and
:here was a 39 percent improvement in days of sunshine observable. The
ıall of smoke was thinning out.

Under the ordinance, factories, homes, steamboats, and locomotives
were required to burn smokeless fuels or install smoke-consuming devices. River boats, which had once fouled the atmosphere, converted to
diesel burners. Apartment houses, churches, hotels, and office buildings
converted to gas fuel. Schools installed stoking equipment and went to
smokeless coal. Private homes, however, were once more exempted for
ı year, chiefly because of the combined opposition of the owners and
ınions involved in soft-coal production.

By 1952 industry, public utilities, railroads, and municipalities in
Allegheny County were at work on a $200,000,000 air pollution control
ırogram. In that year a saving of $26,000,000 was estimated for Pitts-
burgh in cleaning bills alone. Household laundry bills dropped

$5,500,000. Visibility was up 77 percent over 1945. And no one even tried to estimate the benefits to physical and mental health. In the past ten years, the 132 open-hearth furnaces in Pittsburgh have found out how to limit smoke and fly ash, and are now working on methods of screening dust from open-hearth gas.

The Bureau of Air Pollution Control of the Allegheny County Health Department reports that despite record steel production last year and in the first quarter of 1965, the basic downward trend in dustfall continues (though in April of this year the dustfall registered thirty tons per square mile per month, compared with 1962's average of only twenty-seven tons.) And a ten-year report made available by Thomas E. Purcell, chief engineer for the Bureau of Air Pollution Control, shows an almost continuing drop from 1954 through 1964.

Sunny Pittsburgh has come a long way since the 1940s. Most residents and industries now believe in pollution control—at least in the atmosphere. But other troubles persist. Disposal of rubbish, garbage, and waste by municipalities is still a major problem. Sulfur contaminants, carcinogens, and other airborne health hazards are just coming under serious study by health department laboratories, aided by federal grants. And the state is only now preparing legislation to eliminate the privileged status of mines and make acid drainage subject to the same regulations as other industrial waste pouring into Pennsylvania's streams and rivers. The mining industry is, of course, opposed. The cost of avoiding such drainage, it says, would bankrupt every mine.

To anyone who was around thirty years ago, the arguments sound somewhat familiar. But the participants in the struggle have at least one advantage this time: On good days they can enjoy Pittsburgh's fresh air and bright sunshine.

The rebirth of a river

LEE EDSON

Lee Edson is on the staff of Petroleum Today, a quarterly publication
of the Petroleum Institute's Committee on Public Affairs.

A trailer truck carrying drums of cyanide struck a bridge abutment
and careened down the bank of a creek that leads into the Ohio River.
Within minutes radio stations in Louisville, Kentucky, fourteen miles
downstream, were broadcasting the alarm: "Deadly poison on the river
. . moving toward municipal water intake." The frightened public was
advised: "Don't drink city water."

Meanwhile, in a small office in downtown Cincinnati, biologist-
chemist William Klein calmly coordinated efforts to meet the crisis. Investigators were dispatched to the scene of the accident; boats manned
by pollution experts moved out into the river to collect and analyze water
samples; information was exchanged with municipal and industrial representatives.

Two hours after the first report of the accident Klein's office issued
a statement that rapidly dispelled the rising public hysteria. The river
would easily dilute and dissipate the cyanide many hours before the affected water could reach the municipal intake. The Ohio was safe; not
simon-pure, but safe.

Tracking down sources of pollution, though they are normally less
dramatic than cyanide, is part of Klein's job. He is a member of the staff
of the Ohio River Valley Water Sanitation Commission (ORSANCO), an
interstate agency created eighteen years ago to clean up the Ohio and
her tributaries. He and his colleagues are professionally discontented

Reprinted from *Petroleum Today,* Summer 1966.

203

with the condition of the river; at the same time, they are proud of the progress that has been made.

When ORSANCO set up shop in 1948, sewage from the communities along the Ohio was dumped untreated into the river. Today 99. percent of this sewage is first piped into purification plants.

In 1948 the burgeoning industry along the Ohio gave little thought to the vast volume of industrial waste poured forth into the river. Today 90 percent of the 1,700 firms discharging effluents into Ohio Valley streams meet minimum control standards, and many do much better.

The progress achieved under ORSANCO's aegis has had a heady price tag: more than $1 billion for municipal sewage treatment facilities nearly $500 million for industrial treatment. And it is worth noting that about 90 percent of these funds has come from the communities, states and industries involved rather than from the federal government. From its inception, the Ohio River renaissance has been a local matter.

It was during the 1930s that the public first became aware that the "beautiful" Ohio and her nineteen tributaries were turning ugly. The growth of cities and of industry combined to befoul the 981-mile stream. Chemicals and raw sewage blended; the garbage floating down the waterway was described by a morbid mathematician as equal in weight to a dead horse floating downstream every three minutes. One tributary that sometimes ran black with poisoned fish was dubbed the Death River.

During droughts river water collected in hollows behind dams forming stagnant cesspools. Floods then washed these polluted waters into municipal intakes, carrying dysentery and other severe intestinal diseases into thousands of homes.

Voices were raised calling for action to rescue the Ohio and the millions of people who depended on her. Newspapers carried strong editorials protesting the ever-increasing pollution of the waters. But there was little public response.

Then, in 1935, Hudson Biery, a public relations man for the Cincinnati Street Railway Company, entered the lists. He had just finished participating in a successful "Clean Up Cincinnati" campaign sponsored by the local Chamber of Commerce. It seemed to him that a cleanup of the Ohio represented a more pressing need.

Biery's efforts, backed by William F. Wiley, publisher of the Cincinnati *Enquirer* and president of the Chamber of Commerce, led to the formation of an antipollution committee made up of fifty representatives from industry, the professions, and government. They rapidly recognized that the task they had set themselves would not soon nor easily be achieved.

The most immediate roadblock in the way of river reform was the problem of authority. The Ohio meanders through six states (Pennsyl ania, Ohio, West Virginia, Kentucky, Indiana and Illinois), and her waters are replenished from tributaries that originate in New York and Virginia. Unless there was agreement among these states on such matters as water standards and enforcement policies, any cleanup campaign would fail. It seemed that only the federal government was big enough and impartial enough to handle so complex an interstate assignment.

But there were many on the antipollution committee who were strongly opposed to such an answer. They suggested, instead, that the eight states together form a compact to clean up their own mess, with federal assistance where necessary.

Interstate compacts were not unknown at the time, but rarely had a covenant of such scope, involving so many states in so broad a program, been proposed. It met with considerable opposition: some lawyers and governmental authorities thought it might be illegal; some industry leaders thought the compact would "regulate us out of business."

The Cincinnati committee was instrumental in organizing and promoting a survey of the Ohio River basin, under the auspices of the United States Corps of Engineers and the Public Health Service, to identify the many sources of pollution. It was the most elaborate survey of its kind ever conducted, and its results helped swing many area leaders to the side of the compact. In 1936 the members of the committee celebrated their first major victory when Congress voted to permit the states to form a compact. But it was to be twelve years before they had cause for another major celebration.

Several of the individual state legislatures hesitated to ratify the agreement. Virginia, for example, insisted upon first setting up her own pollution control board to enforce antipollution regulations; West Virginia refused to sign the compact until Virginia did. Pennsylvania also had her doubts. Meanwhile World War II intervened and set back negotiations. Finally, on June 30, 1948, in a historic session in Cincinnati, the governors of the eight states affixed their names to the nation's largest interstate treaty to fight river pollution.

Out of this compact came ORSANCO, an agency assigned the task of coordinating cleanup efforts in the 154,000-square-mile area drained by the Ohio and its tributaries. The agency was charged with promulgating control regulations for the area's waterways. Under the compact, ORSANCO was granted some limited powers of enforcement, but the enforcement role rested primarily with the eight states themselves. The agency became chief observer and watchdog of the rivers, constantly testing and analyzing their condition, ferreting out sources of pollution, providing state authorities with the facts. And as a vital corollary to this role, ORSANCO set about engaging and mobilizing public opinion in the valley to gain and hold broad support for the antipollution fight.

The need to clean up the Ohio was recognized in theory by vi
tually all the leaders of the communities and industries involved. Bu
huge sums of money were required. In 1953, for example, the cost of
sewage treatment plant for Pittsburgh was estimated at $85 million, an
at first the administration balked. But the proponents of the plant eve
tually won the day; the new sewage treatment system was completed i
1959.

Some communities proved to be more stubborn. Maysville, Ken
tucky, led by its mayor, a peppery widow named Rebekah Hord, ignore
urgent requests from Kentucky and ORSANCO that a sewage treatmer
facility be constructed. For twelve years she refused to budge from he
position that the town, with its population of 8,400 could not afford th
plant, and she dared the state to put her and the entire community i
jail.

Finally Kentucky went to the courts, and an order was issued finin
the town $1,000 a day until it complied with state orders. So it was tha
a sewage treatment expert was brought in to Maysville and eventually
bond issue was passed. Today Maysville has a modern sewage plant.

While pressing forward to halt the dumping of raw sewage into th
Ohio, ORSANCO and the states of the compact were also moving to tur
back the increasing tides of river-borne industrial waste. The task wa
complex, for each of the hundreds of firms in the river basin worked wit
different materials and thus had different waste disposal problems. Mor
over, the commercial vessels on the Ohio, transporting more materi
than flows through the Panama Canal or the Port of New York, repre
sented a pollution source that was difficult to police.

ORSANCO met the challenge by enlisting the cooperation of th
industries themselves. Committees from oil, chemical, coal, paper, met
finishing, and steel companies were invited to meet with representative
of the agency to develop standards for the treatment of waste materia
and to work out concrete solutions to common problems.

"That invitation did it," says ORSANCO's executive director an
chief engineer, Dr. Edward J. Cleary. "Industry became a cooperatin
and contributing member of the partnership. Out of this relationship ha
come a mutual understanding of why projects should be undertaken an
an exploration of how they might best be done. The industrial committe
program has been one of the most fruitful endeavors ORSANCO ha
undertaken."

A severe and long-standing pollution problem in the Ohio Valle
has been the drainage of sulfuric acid from the coal mines and from th
residue from coal mining operations. Within recent years the coal in
dustry advisory committee has developed and endorsed procedures fo
controlling this drainage, and the acid flow is being gradually reduce

Perhaps the most "visible" of river pollutants is oil, and oil company committeemen have worked closely with ORSANCO to set up
proper safeguards and conduct continuous testing to govern the discharge
of petroleum. New techniques have been devised to isolate and dispose
of particularly troublesome refinery waste substances, some of which are
now buried in mile-deep holes drilled for that purpose.

A leading example of industry cooperation with ORSANCO is the
story of the fight against pollution in the heavily industrialized Kanawha
River Valley, which has been called the Ruhr of America. The great network of plants feeds vast quantities of complex chemical wastes, especially organic materials, into the Kanawha River.

In 1958 ten of these firms joined with the West Virginia Water
Resources Division and the United States Geological Survey in an assay
of pollutants in a sixty-two-mile stretch of the river. Out of this study
emerged a program calling for an expenditure of several million dollars
by the companies to reduce drastically the quantity of waste materials
discharged into the outfalls of the river. The program is well on the way
toward successful completion. One technique that the firms have adopted
involves piping of wastes into huge biological treatment plants where
bacteria act to destroy the organic chemicals. The cleansed liquids are
then released into the river at times specified by state antipollution
agencies.

In order to gauge the progress of such control programs and to provide basin-wide data to be used in formulating waste-treatment requirements, ORSANCO has developed an elaborate system of river-watching.
A network of monitoring stations provides the Cincinnati headquarters
with a continuous record of the mineral and oxygen content of the waters
at widely separated sites. The stations also keep track of the water temperature and flow, vital factors in the ability of a stream to cleanse itself.

Seventeen of the monitoring stations are operated by managers of
municipal and private water-treatment plants; eleven are run by the
United States Geological Survey; and thirteen are electronic sentinels
that automatically relay hourly reports by long-distance telephone lines
to Cincinnati.

ORSANCO also draws upon the services of the Weather Bureau for
daily reports on the rate and quantity of river flow. By combining this
information with its monitor reports, and feeding the total into a com-
puter, the agency can rapidly evaluate the effects of a "spill" that jeopard-
izes water use.

The achievements of ORSANCO in coordinating state, municipal,
and industry action or in monitoring the waters would go for nought,
however, if the agency and its program lacked public support. In this
connection Director Cleary frequently quotes Abraham Lincoln: "Public

sentiment is everything. With public sentiment nothing can fail; without it nothing can succeed." The Ohio River cleanup has succeeded to an important degree because of public backing.

Cleary divides the informational and educational activities of the agency into two broad categories: the "shotgun technique," intended for large audiences; the "rifleshot technique," used for specific targets.

For the general public the agency has developed a campaign keyed to the individual citizen's stake in clean water. A typical slogan: "Clean Waters Protect Your Health—Protect Your Job—Protect Your Happiness." The antipollution message in its shotgun form is broadcast in newspaper and magazine articles, on radio and television, in exhibits and brochures and films.

The targets for the rifleshot approach are particular communities and industries. Thus staff members have helped to organize citizens' committees and publicity campaigns in dozens of communities to build public support for sewage treatment bond issues.

Out of such activities has come a goodly share of the improvements made along the Ohio since 1948. ORSANCO has blended the skills of mediator, engineer, scientist, and public affairs expert to lead the crusade for clean waters. And the record of achievement is impressive, though it is by no means complete.

There remain a minority of companies that have failed to assume their responsibilities for improving the rivers. Operators of pleasure boats whose numbers have increased with the upgrading of the waters, too often dump litter and sewage. And there are problems raised by new technologies; atomic energy, for example.

Hot water poured into a river is a form of pollution, and the river water used as coolant for nuclear reactors, as well as for some other industrial purposes, is returned to the river at a higher temperature. Heated water can drive out oxygen, thus killing fish; it also impedes the growth of microorganisms that destroy some kinds of pollutants. In addition, it may accelerate chemical reactions in the rivers, producing toxic materials out of normally harmless ingredients.

The ORSANCO river-watchers are keeping their eyes on the Ohio's temperature. Should it rise too high, local industries would need to reduce their operations until the river returned to normal.

When the eight-state compact was signed in Cincinnati in 1948 there were some observers who commented that it represented nothing more than "pious hopes and paper dreams." Yet out of these hopes and dreams there has emerged a program that has won widespread recognition, including the top award of 1963 from the American Society of Civil Engineers.

THE SOVIET PARALLEL

The convergence of environmental disruption

MARSHALL I. GOLDMAN

In our concern with pollution in the United States, we are sometimes led to believe that pollution is an inevitable by-product of capitalism. A study of the Soviet Union shows, however, that environmental disruption, in all the forms we are so familiar with here, also exist there. The author is working on a longer monograph devoted entirely to the question of environmental disruption in the Soviet Union.

By now it is a familiar story: rivers that blaze with fire, smog that suffocates cities, streams that vomit dead fish, oil slicks that blacken seacoasts, prized beaches that vanish in the waves, and lakes that evaporate and die a slow smelly death. What makes it unfamiliar is that this is a description not only of the United States but also of the Soviet Union.

Most conservationists and social critics are unaware that the USSR has environmental disruption that is as extensive and severe as ours. Most of us have been so distressed by our own environmental disruption that we lack the emotional energy to worry about anyone else's difficulties. Yet before we can find a solution to the environmental disruption in our own country, it is necessary to explain why it is that a socialist or communist country like the USSR finds itself abusing the environment in the same way, and to the same degree, that we abuse it. This is especially important for those who have come to believe as basic doctrine that it is capitalism and private greed that are the root causes of environmental disruption. Undoubtedly private enterprise and the profit motive account for a good portion of the environmental disruption that we encounter in this country. However, a study of pollution in the Soviet Union suggests

Reprinted with permission from *Science*, Vol. 170, 37–42, 2 October 1970, © 1970 by the American Association for the Advancement of Science.

that abolishing private proverty will not necessarily mean an end to en
vironmental disruption. In some ways, state ownership of the country'
productive resources may actually exacerbate rather than ameliorate the
situation.

THE PUBLIC GOOD

That environmental disruption is a serious matter in the Soviet Union
usually comes as a surprise not only to most radical critics of pollution in
the West but also to many Russians. It has been assumed that, if all the
factories in the society were state-owned, the state would insure that the
broader interests of the general public would be protected. Each factory
would be expected to bear the full costs and consequences of its opera
tion. No factory would be allowed to take a particular action if it mean
that the public would suffer or would have to bear the expense. In other
words, the factory would not only have to pay for its *private costs*, such
as expenses for labor and raw materials; it would also have to pay for its
social costs, such as the cost of eliminating the air and water pollution i
had caused. It was argued that, since the industry was state-run, includ
ing both types of costs would not be difficult. At least that was what was
assumed.

Soviet officials continue today to make such assumptions. B. V. Pe
trovsky, the Soviet Minister of Public Health, finds environmental dis
ruption in a capitalist society perfectly understandable: "the capitalist
system by its very essence is incapable of taking radical measures to en
sure the efficient conservation of nature." By implication he assumes that
the Soviet Union can take such measures. Therefore it must be somewhat
embarrassing for Nikolai Popov, an editor of *Soviet Life,* to have to ask
"Why, in a socialist country, whose constitution explicitly says the public
interest may not be ignored with impunity, are industry executives per
mitted to break the laws protecting nature?"

Behind Popov's question is a chronicle of environmental disruption
that is as serious as almost any that exists in the world. Of course in a
country as large as the USSR there are many places that have been
spared man's disruptive incursions. But, as the population grows in num
bers and mobility, such areas become fewer and fewer. Moreover, as in
the United States, the most idyllic sites are the very ones that tend to
attract the Soviet population.

Just because human beings intrude on an area, it does not neces
sarily follow that the area's resources will be abused. Certainly the pres
ence of human beings means some alteration in the previous ecological
balance, and in some cases there may be severe damage, but the change

need not be a serious one. Nonetheless, many of the changes that have taken place in the Soviet Union have been major ones. As a result, the quality of the air, water, and land resources has been adversely affected.

WATER

Comparing pollution in the United States and in the USSR is something like a game. Any depressing story that can be told about an incident in the United States can be matched by a horror story from the USSR. For example, there have been hundreds of fish-kill incidents in both countries. Rivers and lakes from Maine to California have had such incidents. In the USSR, effluent from the Chernorechensk Chemical Plant near Dzerzhinsk killed almost all the fish life in the Oka River in 1965 because of uncontrolled dumping. Factories along major rivers such as the Volga, Ob, Yenesei, Ural, and Northern Dvina have committed similar offenses, and these rivers are considered to be highly polluted. There is not one river in the Ukraine whose natural state has been preserved.[1] The Molognaia River in the Ukraine and many other rivers throughout the country are officially reported as dead. How dangerous this can be is illustrated by what happened in Sverdlovsk in 1965. A careless smoker threw his cigarette into the Iset River and, like the Cuyahoga in Cleveland, the Iset caught fire.

Sixty-five percent of all the factories in the largest Soviet republic, the Russian Soviet Federated Socialist Republic (RSFSR), discharge their waste without bothering to clean it up.[2] But factories are not the only ones responsible for the poor quality of the water. Mines, oil wells, and ships freely dump their waste and ballast into the nearest body of water. Added to this industrial waste is the sewage of many Russian cities. Large cities like Moscow and Leningrad are struggling valiantly, like New York and Chicago, to treat their waste, but many municipalities are hopelessly behind in their efforts to do the job properly. Only six out of the twenty main cities in Moldavia have a sewer system, and only two of those cities make any effort to treat their sewage.[3] Similarly, only 40 percent of the cities and suburbs in the RSFSR have any equipment for treating their sewage. For that matter, according to the last completed census, taken in 1960, only 35 percent of all the housing units in urban areas are served by a sewer system.[4]

[1]*Rabochaia Gaz.* 1967, 4 (15 Dec. 1967).
[2]*Ekon. Gaz.* 1967, No. 4, 37 (1967).
[3]*Sovet. Moldaviia* 1969, 2 (1 June 1969).
[4]V. G. Kriazhev, *Vnerabochee Vremia i Sphera Obslyzhivaniia* (Ekonomika, Moscow, 1966), p. 130.

Conditions are even more primitive in the countryside. Often thi adversely affects the well-water and groundwater supplies, especially i areas of heavy population concentration. Under the circumstances it i not surprising to find that major cities like Vladimir, Orenburg, and Voro nezh do not have adequate supplies of drinking water. In one instanc reported in *Pravda,* a lead and zinc ore enriching plant was built in 196 and allowed to dump its wastes in the Fragdon River, even though th river was the sole source of water for about 40 kilometers along its route As a result the water became contaminated and many people were sim ply left without anything to drink.

Even when there are supplies of pure water, many homes through out the country are not provided with running water. This was true o 62 percent of the urban residences in the USSR in 1960.[5] The Russian often try to explain this by pointing to the devastation they suffered dur ing World War II. Still it is something of a shock, twenty-five years afte the war, to walk along one of the more fashionable streets in Kharkov the fifth largest city in the USSR, and see many of the area's resident with a yoke across their shoulders, carrying two buckets of water. Th scene can be duplicated in almost any other city in the USSR.

Again, the Soviet Union, like the United States, has had trouble no only with its rivers but with its larger bodies of water. As on Cape Co and along the California coast, oil from slicks has coated the shores of th Baltic, Black, and Caspian seas. Refineries and tankers have been espe cially lax in their choice of oil-disposal procedures.

Occasionally it is not only the quality but the quantity of the wate that causes concern. The Aral and Caspian seas have been gradually dis appearing. Because both seas are in arid regions, large quantities of thei water have been diverted for crop irrigation. Moreover, many dams anc reservoirs have been built on the rivers that supply both seas for th generation of electric power. As a result of such activities, the Aral Sea began to disappear. From 1961 to 1969 its surface dropped 1 to 3 meters Since the average depth of the sea is only about 20 to 30 meters, som Russian authorities fear that, at the current rate of shrinkage, by the tur of the century the sea will be nothing but a salt marsh.[5]

Similarly, during the past twenty years the level of the Caspian Sea has fallen almost 2.5 meters. This has drastically affected the sea's fish population. Many of the best spawning areas have turned into dry land For the sturgeon, one of the most important fish in the Caspian, this has meant the elimination of one-third of the spawning area. The combinec effect of the oil on the sea and the smaller spawning area reduced th fish catch in the Caspian from 1,180,400 centners in 1942 to 586,300 cent-

[5]*Soviet News* 1970, 6 (7 Apr. 1970).

ners in 1966. Food fanciers are worried not so much about the sturgeon as about the caviar that the sturgeon produces. The output of caviar has fallen even more drastically than the sea level—a concern not only for the Russian consumers of caviar but for foreigners. Caviar had been a major earner of foreign exchange. Conditions have become so serious that the Russians have now begun to experiment with the production of artificial caviar.

The disruption of natural life has had some serious ecological side effects. Near Ashkhabad a fish called the belyi amur also began to disappear. As a consequence, the mosquito population, which had been held in check by the belyi amur, grew in the newly formed swamps where once the sea had been. In turn, the mosquitoes began to transmit malaria.[6]

Perhaps the best known example of the misuse of water resources in the USSR has been what happened to Lake Baikal. This magnificent lake is estimated to be over 20 million years old. There are over twelve hundred species of living organisms in the lake, including freshwater seals and seven hundred other organisms that are found in few or no other places in the world. It is one of the largest and deepest freshwater lakes on earth, over 1.5 kilometers deep in some areas.[7] It is five times as deep as Lake Superior and contains twice the volume of water. In fact, Lake Baikal holds almost one-fortieth of all the world's fresh water. The water is low in salt content and is highly transparent; one can see as far as 36 meters under water.[8]

In 1966, first one and then another paper and pulp mill appeared on Lake Baikal's shores. Immediately limnologists and conservationists protested this assault on an international treasure. Nonetheless, new homes were built in the vicinity of the paper and pulp mills, and the plant at the nearby town of Baikalsk began to dump 60 million cubic meters of effluent a year into the lake. A specially designed treatment plant had been erected in the hope that it would maintain the purity of the lake. Given the unique quality of the water, however, it soon became apparent that almost no treatment plant would be good enough. Even though the processed water is drinkable, it still has a yellowish tinge and a barely perceptible odor. As might be expected, a few months after this effluent had been discharged into the lake, the Limnological Institute reported that animal and plant life had decreased by one-third to one-half in the zone where the sewage was being discharged.

Several limnologists have argued that the only effective way to prevent the mill's effluent from damaging the lake is to keep it out of the

[6]*Turkm. Iskra* 1969, 3 (16 Sept. 1969).
[7]O. Volkov, *Soviet Life* 1966, 6 (Aug. 1966).
[8]L. Rossolimo, *Baikal* (Nauka, Moscow, 1966), p. 91.

lake entirely. They suggest that this can be done if a 67-kilometer sewage conduit is built over the mountains to the Irkut River, which does not flow into the lake. So far the Ministry of Paper and Pulp Industries has strongly opposed this, since it would cost close to $40 million to build such a bypass. They argue that they have already spent a large sum on preventing pollution. Part of their lack of enthusiasm for any further change may also be explained by the fact that they have only had to pay fines of $55 for each violation. It has been cheaper to pay the fines than to worry about a substantial cleanup operation.

Amid continuing complaints, the second paper and pulp mill, at Kamensk, was told that it must build and test its treatment plant before production of paper and pulp would be allowed. Moreover, the lake and its entire drainage basin have been declared a "protected zone," which means that in the future all timber cutting and plant operations are to be strictly regulated. Many critics, however, doubt the effectiveness of such orders. As far back as 1960, similar regulations were issued for Lake Baikal and its timber, without much result. In addition, the Ministry of Pulp and Paper Industries has plans for constructing yet more paper and pulp mills along the shores of Lake Baikal and is lobbying for funds to build them.

Many ecologists fear that, even if no more paper mills are built, the damage may already have been done. The construction of the mills and towns necessitated the cutting of trees near the shoreline, which inevitably increased the flow of silt into the lake and its feeder streams. Furthermore, instead of being shipped by rail, as was originally promised, the logs are rafted on the water to the mill for processing. Unfortunately about 10 percent of these logs sink to the lake bottom in transit. Not only does this cut off the feeding and breeding grounds on the bottom of the lake but the logs consume the lake's oxygen, which again reduces its purity.

There are those who see even more dire consequences from the exploitation of the timber around the lake. The Gobi Desert is just over the border in Mongolia. The cutting of the trees and the intrusion of machinery into the wooded areas has destroyed an important soil stabilizer. Many scientists report that the dunes have already started to move, and some fear that the Gobi Desert will sweep into Siberia and destroy the taiga and the lake.

AIR

The misuse of air resources in the USSR is not very different from the misuse of water. Despite the fact that the Russians at present produce less than one-tenth the number of cars each year that we produce in the United States, most Soviet cities have air pollution. It can be quite serious, especially when the city is situated in a valley or a hilly region. In the hilly cities of Armenia, the established health norms for carbon monoxide are often exceeded. Similarly Magnitogorsk, Alma Ata, and Chelyabinsk, with their metallurgical industries, frequently have a dark blue cap over them. Like Los Angeles, Tbilisi, the capital of the Republic of Georgia, has smog almost 6 months of the year. Nor is air pollution limited to hilly regions. Leningrad has 40 percent fewer clear daylight hours than the nearby town of Pavlovsk.[9]

Of all the factories that emit harmful wastes through their stacks, only 14 percent were reported in 1968 to have fully equipped air-cleaning devices. Another 26 percent had some treatment equipment. Even so, there are frequent complaints that such equipment is either operating improperly or of no use. There have been several reported instances of factories' spewing lead into the air.[10] In other cases, especially in Sverdlovsk and Magnitogorsk, public health officials ordered the closing of factories and boilers. Nevertheless, there are periodic complaints that some public health officials have yielded to the pleadings and pressures of factory directors and have agreed to keep the plants open "on a temporary basis."

One particularly poignant instance of air pollution is occurring outside the historic city of Tula. Not far away is the site of Leo Tolstoy's former summer estate, Yasnaya Polyana, now an internationally known tourist attraction with lovely grounds and a museum. Due to some inexcusable oversight, a small coal-gasification plant was built within view of Yasnaya Polyana in 1955. In 1960 the plant was expanded as it began to produce fertilizer and other chemicals. Now known as the Shchkino Chemical Complex, the plant has over six thousand employees and produces a whole range of chemicals, including formaldehyde and synthetic fibers. Unfortunately the prevailing winds from this extensive complex blow across the street onto the magnificent forests at Yasnaya Polyana. As a result, a prime oak forest is reported near extinction and a pine forest is similarly affected.

LAND

As in other nations of the world, environmental disruption in the USSR is not limited to air and water. For example, the Black Sea coast in the

[9]I. Petrov, *Kommunist* 1969, No. 11, 74 (1969).
[10]*Rabochaia Gaz.* 1969, 4 (27 June 1969); *Ekon. Gaz.* 1968, No. 4, 40 (1968); *Lit. Gaz.* 1967, No. 32, 10 (1967).

Soviet Republic of Georgia is disappearing. Since this is a particularly desirable resort area, a good deal of concern has been expressed over what is happening. At some places the sea has moved as much as 40 meters inland. Near the resort area of Adler, hospitals, resort hotels, and (of all things) the beach sanitarium of the Ministry of Defense collapsed as the shoreline gave way. Particular fears that the mainline railway will also be washed away shortly have been expressed.

New Yorkers who vacation on Fire Island have had comparable difficulties, but the cause of the erosion in the USSR is unique. Excessive construction has loosened the soil (as at Fire Island) and accelerated the process of erosion. But, in addition, much of the Black Sea area has been simply hauled away by contractors. One contractor realized that the pebbles and sand on the riviera-type beach were a cheap source of gravel. Soon many contractors were taking advantage of nature's blessings. As a result, as much as 120,000 cubic meters a year of beach material has been hauled away. Unfortunately the natural process by which those pebbles are replaced was disrupted when the state came along and built a network of dams and reservoirs on the stream feeding into the sea. This provided a source of power and water but it stopped the natural flow of pebbles and sand to the seacoast. Without the pebbles, there is little to cushion the enormous power of the waves as they crash against the coast and erode the shoreline.

In an effort to curb the erosion, orders have been issued to prevent the construction of any more buildings within 3 kilometers of the shore. Concrete piers have also been constructed to absorb the impact of the waves, and efforts are being made to haul gravel material from the inland mountains to replace that which has been taken from the seacoast. Still the contractors are disregarding the orders—they continue to haul away the pebbles and sand, and the seacoast continues to disappear.

Nor is the Black Sea coast the only instance of such disregard for the forces of nature. High in the Caucasus is the popular health resort and spa of Kislovodsk. Surrounded on three sides by a protective semicircle of mountains which keep out the cold winds of winter, the resort has long been noted for its unique climate and fresh mountain air. Whereas Kislovodsk used to have 311 days of sun a year, Piatagorsk on the other side of the mountain had only 122.[11] Then, shortly after World War II, an official of the Ministry of Railroad sought to increase the volume of railroad freight in the area. He arranged for the construction of a lime kiln in the nearby village of Podkumok. With time, pressure mounted to increase the processing of lime, so that now there are eight

[11]*Izv.* 1966, 5 (3 July 1966).

kilns in operation. As the manager of the lime kiln operation and railroad officials continued to "fulfill their ever-increasing plan" in the name of "socialist competition," the mountain barrier protecting Kislovodsk from the northern winds and smoke of the lime kilns has been gradually chopped away. Consequently, Kislovodsk has almost been transformed into an ordinary industrial city. The dust in the air now exceeds by 50 percent the norm for a *nonresort* city.

Much as some of our ecologists have been warning that we are on the verge of some fundamental disruptions of nature, so the Russians have their prophets of catastrophe. Several geographers and scientists have become especially concerned about the network of hydroelectric stations and irrigation reservoirs and canals that have been built with great fanfare across the country. They are now beginning to find that such projects have had several unanticipated side effects. For example, because the irrigation canals have not been lined, there has been considerable seepage of water. The seepage from the canals and an overenthusiastic use of water for irrigation has caused a rise in the water table in many areas. This has facilitated salination of the soil, especially in dry areas. Similarly, the damming of water bodies apparently has disrupted the addition of water to underground water reserves. There is concern that age-old sources of drinking water may gradually disappear. Finally, it is feared that the reduction of old water surfaces and the formation of new ones has radically altered and increased the amount of water evaporation in the area in question. There is evidence that this has brought about a restructuring of old climate and moisture patterns.[12] This may mean the formation of new deserts in the area. More worrisome is the possibility of an extension of the ice cap. If enough of Russia's northward-flowing rivers are diverted for irrigation purposes to the arid south, this will deprive the Arctic Ocean of the warmer waters it receives from these rivers. Some scientist critics also warn that reversing the flow of some of the world's rivers in this way will have disruptive effects on the rotation of the earth.

REASONS FOR POLLUTION

Because the relative impact of environmental disruption is a difficult thing to measure, it is somewhat meaningless to say that the Russians are more affected than we are, or vice versa. But what should be of interest is an attempt to ascertain why it is that pollution exists in a state-owned,

[12]*Soviet News* 1969, 105 (11 March 1969).

centrally planned economy like that of the Soviet Union. Despite the fact that our economies differ, many if not all of the usual economic explanations for pollution in the non-Communist world also hold for the Soviet Union. The Russians, too, have been unable to adjust their accounting system so that each enterprise pays not only its direct costs of production for labor, raw materials, and equipment but also its social costs of production arising from such byproducts as dirty air and water. If the factory were charged for these social costs and had to take them into account when trying to make a profit on its operations, presumably factories would throw off less waste and would reuse or recycle their air and water. However, the precise social cost of such waste is difficult to measure and allocate under the best of circumstances, be it in the United States or the USSR. (In the Ruhr Valley in Germany, industries and municipalities are charged for the water they consume and discharge, but their system has shortcomings.)

In addition, almost everyone in the world regards air and water as free goods. Thus, even if it were always technologically feasible, it would still be awkward ideologically to charge for something that "belongs to everyone," particularly in a Communist society. For a variety of reasons, therefore, air and water in the USSR are treated as free or undervalued goods. When anything is free, there is a tendency to consume it without regard for future consequences. But with water and air, as with free love, there is a limit to the amount available to be consumed, and after a time there is the risk of exhaustion. We saw an illustration of this principle in the use of water for irrigation. Since water was treated virtually as a free good, the Russians did not care how much water they lost through unlined canals or how much water they used to irrigate the soil.

Similarly, the Russians have not been able to create clear lines of authority and responsibility for enforcing pollution-control regulations. As in the United States, various Russian agencies, from the Ministry of Agriculture to the Ministry of Public Health, have some but not ultimate say in coping with the problem. Frequently when an agency does attempt to enforce a law, the polluter will deliberately choose to break the law. As we saw at Lake Baikal, this is especially tempting when the penalty for breaking the law is only $55 a time, while the cost of eliminating the effluent may be in the millions of dollars.

The Russians also have to contend with an increase in population growth and the concentration of much of this increase in urban areas. In addition, this larger population has been the beneficiary of an increase in the quantity and complexity of production that accompanies industrialization. As a result, not only is each individual in the Soviet Union, as in the United States, provided with more goods to consume, but the resulting products, such as plastics and detergents, are more exotic and less easily disposed of than goods of an earlier, less complicated age.

Like their fellow inhabitants of the world, the Russians have to contend with something even more ominous than the Malthusian Principle. Malthus observed that the population increased at a geometric rate but that food production grew at only an arithmetic rate. If he really wants to be dismal, the economist of today has more to worry about. It is true that the population seems to be increasing at accelerated rates, but, whereas food production at least continues to increase, our air, water, and soil supplies are relatively constant. They can be renewed, just as crops can be replanted, but, for the most part, they cannot be expanded. In the long run, this "Doomsday Principle" may prove to be of more consequence than the Malthusian doctrine. With time and pollution we may simply run out of fresh air and water. Then, if the damage is not irreversible, a portion of the population will be eliminated and those who remain will exist until there is a shortage once again or until the air, water, and soil are irretrievably poisoned.

INCENTIVES TO POLLUTE UNDER SOCIALISM

In addition to the factors which confront all the people of the earth, regardless of their social or economic system, there are some reasons for polluting which seem to be peculiar to a socialist country such as the Soviet Union in its present state of economic development. First of all, state officials in the Soviet Union are judged almost entirely by how much they are able to increase their region's economic growth. Thus, government officials are not likely to be promoted if they decide to act as impartial referees between contending factions on questions of pollution. State officials identify with the polluters, not the conservationists, because the polluters will increase economic growth and the prosperity of the region while the antipolluters want to divert resources away from increased production. There is almost a political as well as an economic imperative to devour idle resources. The limnologists at Lake Baikal fear no one so much as the voracious Gosplan (State Planning) officials and their allies in the regional government offices. These officials do not have to face a voting constituency which might reflect the conservation point of view, such as the League of Women Voters or the Sierra Club in this country. It is true that there are outspoken conservationists in the USSR who are often supported by the Soviet press, but for the most part they do not have a vote. Thus the lime smelters continued to smoke away behind the resort area of Kislovodsk even though critics in *Izvestiia*,

Literaturnaya Gazeta, Sovetskaia Rossiia, Trud, and *Krokodil* protested long and loud.

At one time state governments in our country often reflected similar one-sidedness. Maine, for example, was often cited as an area where industry did what it wanted to do to nature. Now, as the conservationist voting bloc has grown in size, the Maine state government finds itself acting as referee. Accordingly it has passed a far-reaching law which regulates the location and operation of all new industry. Failure to have voted for such legislation may have meant defeat at the polls for many politicians. No such device for transmitting voting pressure exists at present in the USSR.

Second, industrialization has come relatively recently to the USSR and so the Russians continue to emphasize the increase in production. Pollution control generally appears to be non-productive, and there is usually resistance to the diversion of resources from productive to nonproductive purposes. This is even reflected in the words used to describe the various choices. "Conserve" generally seems to stand in opposition to "produce."

Third, until July 1967, all raw materials in the ground were treated by the Russians as free goods. As a result, whenever the mine operator or oil driller had exploited the most accessible oil and ore, he moved on to a new site where the average variable costs were lower. This has resulted in very low recovery rates and the discarding of large quantities of salvageable materials, which increase the amount of waste to be disposed of.

Fourth, as we have seen, it is as hard for the Russians as it is for us to include social costs in factory-pricing calculations. However, not only do they have to worry about social cost accounting, they also are unable to reflect all the private cost considerations. Because there is no private ownership of land, there are no private property owners to protest the abuse of various resources. Occasionally it does happen that a private property owner in the United States calculates that his private benefits from selling his land for use in some new disruptive use is *not* greater than the private cost he would bear as a result of not being able to use the land any more. So he retains the land as it is. The lack of such private property holders or resort owners and of such a calculation seems to be the major reason why erosion is destroying the Black Sea coast. There is no one who can lay claim to the pebbles on the shore front, and so they are free to anyone who wants to cart them away. Of course private landowners do often decide to sell their land, especially if the new use is to be for oil exploitation rather than pebble exploitation. Then the private benefits to the former owner are high and the social costs are ignored, as always. The Russians, however, under their existing system, now only have to worry about accounting for social costs, they lack the first line of protection that would come from balancing private costs and private benefits.

Fifth, economic growth in the USSR has been even more unbalanced, and in some cases more one-sided, than in the United States. Thus, occasionally change takes place so rapidly and on such a massive scale in a state-run economy that there is no time to reflect on all the consequences. In the early 1960s, Khrushchev decided that the Soviet Union needed a large chemical industry. All at once chemical plants began to spring up or expand all over the country. In their anxiety to fulfill their targets for new plant construction, few if any of the planners were able to devote much attention to the disruptive effects on the environment that such plants might have. We saw one result at Yasnaya Polyana. In fact, the power of the state to make fundamental changes may be so great that irreversible changes may frequently be inflicted on the environment without anyone's realizing what is happening until it is too late. This seems to be the best explanation of the meteorological disruption that is taking place in Siberia. It is easier for an all-powerful organism like the state than for a group of private entrepreneurs to build the reservoirs and reverse the rivers. Private enterprises can cause their own havoc, as our dust bowl experience or our use of certain pesticides or sedatives indicates, but in the absence of private business or property interests the state's powers can be much more far-reaching in scope. In an age of rampant technology, where the consequences of one's actions are not always fully anticipated, even well-intentioned programs can have disastrous effects on the environmental status quo.

ADVANTAGES OF A SOCIALIST SYSTEM

Amidst all these problems, there are some things the Russians do very well. For example, the Russians have the power to prevent the production of various products. Thus, the Soviet Union is the only country in the world that does not put ethyl lead in most of the gasoline it produces. This may be due to technical lag as much as to considerations of health, but the result is considerably more lead-free gasoline. Similarly, the Russians have not permitted as much emphasis on consumer-goods production as we have in the West. Consequently there is less waste to discard. Russian consumers may be somewhat less enthusiastic about this than the ecologists and conservationists, but in the USSR there are no disposable bottles or disposable diapers to worry about. It also happens that, because labor costs are low relative to the price of goods, more emphasis is placed on prolonging the life of various products. In other words it is worthwhile to use labor to pick up bottles and collect junk. No one would intentionally abandon his car on a Moscow street, as 70,000 people did in New York City in 1970. Even if a Russian car is twenty years old, it is still valuable. Because of the price relationships that exist in the USSR, the junkman can still make a profit. This facilitates the recycling process, which ecologists tell us is the ultimate solution to environmental disruption.

It should also be remembered that, while not all Russian laws are observed, the Russians do have an effective law enforcement system which they have periodically brought to bear in the past. Similarly, they have the power to set aside land for use as natural preserves. The lack of private land ownership makes this a much easier process to implement than in the United States. As of 1969, the Soviet Government had set aside eighty such preserves, encompassing nearly 65,000 square kilometers.

Again because they own all the utilities as well as most of the buildings, the Russians have stressed the installation of centrally supplied steam. Thus, heating and hot water are provided by central stations, and this makes possible more efficient combustion and better smoke control than would be achieved if each building were to provide heat and hot water for itself. Although some American cities have similar systems, this approach is something we should know more about.

In sum, if the study of environmental disruption in the Soviet Union demonstrates anything, it shows that not private enterprise but industrialization is the primary cause of environmental disruption. This suggests that state ownership of all the productive resources is not a cure-all. The replacement of private greed by public greed is not much of an improvement. Currently the proposals for the solution of environmental disruption seem to be no more advanced in the USSR than they are in the United States. One thing does seem clear, however, and that is that, unless the Russians change their ways, there seems little reason to believe that a strong centralized and planned economy has any notable advantages over other economic systems in solving environmental disruption.

PART 5

CONCLUSION

Where do we go from here?

SENATOR EDMUND S. MUSKIE

Asked to address a Coastal Conference in South Portland, Maine, by the Casco Bay Island Development Association, Senator Muskie took a regional and state issue and put it in the framework of the general problem of environmental disruption. It is an excellent statement of how various interest groups are affected whether they want to be or not. Although Senator Muskie is addressing himself primarily to Maine's problems, what he has to say is applicable to any state or nation in the world.

As I think of the issues that have brought you together, I can't help but think that we really ought to give thanks to the crises which bring us together to consider important problems from time to time—problems which all too often are not discussed at all until crisis strikes and finds us ill-prepared to deal with the problem. Major controversies with all of the decisive impact that they have do have a beneficial effect of alerting us to problems which are growing, almost unseen in our midst. So controversy is rather in the nature of an early warning system—like squalls before the storm—to tell us of greater danger to come. And I think that this is the way we ought to look at this controversy and the opportunity it creates for us to look ahead more perceptibly with greater understanding and, perhaps after full dialogue, with greater wisdom.

I was glad to learn that the Casco Bay Island Development Association has decided to treat the controversy over Long Island as a warning signal. Your decision makes sense in terms of the Portland harbor problem related to existing and potential oil pollution. It also gives us an opportunity to focus more sharply on the steps we need to take to protect the entire coast of Maine from a variety of threats.

Reprinted with permission of Senator Edmund S. Muskie, from a speech at the Casco Bay Coastal Conference, July 19, 1969.

Looking back on the debate of the zoning application for Long Island, I'm sure that most of us could generate suggestions on what other courses might have been followed by the city of Portland, by the Portland metropolitan region, and by the state. The fact is, however, that none of these jurisdictions was prepared for all the questions involved in the desire of a company to convert a former naval oil depot into an active and potentially large commercial oil depot and transfer point. This kind of purpose is certainly consistent to the free enterprise system. It interested the company in pursuing its objectives and its purposes. But the fact is that the public sector is unprepared to deal with the problems created.

Considering the absence of a comprehensive plan for the metropolitan area of Portland, including the islands of the Bay; considering the variety of interests concerned with the future uses of the land and water resources of this region; considering the uncertainty over the impact of the King Resources project on Casco Bay, Portland harbor, and the communities in the area—I think the city can take pride in the reasoned and constructive approach which has been followed by and large by its officials and agencies.

Now that isn't to say that the battle is over. On the narrow question of oil pollution, the city of Portland and the adjacent communities must insure that adequate public authority exists to require maximum protection against oil discharges from vessels and shore installations to offshore installations and transfer operations.

We must insist on the latest and best cleanup equipment and procedures in the event of discharge. And we must require proof of financial responsibility and assumption of that responsibility by potential dischargers to clean up or to pay the cost of cleaning up discharge and to reimburse injured parties for damages from such discharge. Portland is in a position to be tough on these counts. I think Portland must be tough and I think no enterprise can be justified unless it is prepared to accept tough public policies in these areas.

I know it's been said many times in recent years that one of the benefits we can derive from Maine's relatively slow economic growth and development is that we're in a position to learn from the mistakes made by areas which grew faster and developed earlier than we. During the Industrial Revolution there was scant awareness of the safeguards that could be taken to avoid the unfavorable impact of industrialization upon nature and upon our environment. But we are in a position to take advantage of the mistakes that have been made, and we should, to the maximum that is possible with existing technology, and an enlightened public policy which the public has given every evidence it is willing to support. Only agencies adequately backed both by statutory requirements that empower them to issue and enforce meaningful regulations and by a staff with sufficient qualified personnel to implement their authority can be effective, and they should not focus just on the proposed Long Island project. Portland Harbor and Casco Bay already have serious oil pollution problems which are potentially much worse with the King Resources proposals. We ought to take advantage of the present critical decision-making problem which we face to deal with the total problem.

And may I add this—always when tougher public policy is proposed at the federal level, the state level, or the local level, in areas such as this, we hear testimony from companies affected, from the industrial sector, to determine if their policy is consistent with the public interest and if it is their intent to give due consideration to the safeguards that must be taken.

But with all respect to that kind of good citizenship which we do find in the private sector, the requirements I am talking about cannot be left to the good will and the promises of the companies themselves. Public protection is a civic responsibility but the ultimate authority for its implementation must always rest in public hands.

In a Senate committee on public works at the present time, we are working towards what we hope will be effective national legislation covering oil pollution. Most of the controversy involved in our deliberation has been hidden from public view because it has been going on for almost two months now in executive session. As author of the bill and chairman of the committee, I'm pressing for tough, new, and unprecedented public policy—the concept of absolute liability for cleanup of oil spills—and the votes are close. The last one was 8 to 6. We may lose the next one by 8 to 6.

I cite this for one reason. To illustrate the fact that notwithstanding the existence of a great sensitivity on the part of those involved in the oil industry to public views and to public pressure, nevertheless, the industry still looks for the potential way out of a future liability. And only the public agency, whether it's a legislative body or an agency empowered by a legislative body, can be reliable—and not always is it totally reliable—but it is the most likely reliable safeguard for the public interest in these instances.

Now the legislation we're considering will best serve as support for local and regional efforts. More adequate state legislation is required even if the national legislation takes the shape I'd like it to take. So we need state legislation for the entire coast of Maine but unfortunately for your purposes such legislation may be as much as two years away. Portland and its neighboring coastal communities need to act now, for your own

protection and as a model for other local and state programs throughou
our state and throughout the country.

You have heard and you will hear more about the details of pollu
tion control requirements for the bay and the harbor but one fact shoulc
be very clear to you—enactment of local ordinances or establishment o
control agencies will not provide absolute protection against oil spills
There are peculiar and serious hazards in the bay and the harbor whicl
ought not to be overlooked, both for navigation and for the transfer o
oil. The technology of control at sea is still in its infancy. And so or
dinances should not be written just in terms of existing technology. The
should give adequate authority to the enforcement agency to revise rule
and regulations to take advantage of technological developments and t
press operators for ever higher standards of support which we can expec
out of the technology which is sure to emerge under the pressure of pub
lic concern. So make your ordinance a living ordinance, one which ca
move with the times to give you ever greater protection.

One of the main deficiencies in Maine's existing water pollution con
trol statute is the absence of this kind of adequate administrative au
thority to meet changing needs and conditions. It has been our habit t
write detailed description on water quality in statutes. Well, that ma
satisfy short-term desires, especially the desires of those who are bein
regulated but it retards long-term capacity to upgrade environmenta
quality. Do not be trapped by this in your local ordinance.

What's going on in outer space this weekend surely demonstrate
to all of us the tremendous capacity for development in the field of tech
nology which Americans can produce under pressure. And we've got tha
kind of pressure in the environmental quality field so let's gear our publi
policy, our ordinances, our statutes, our public agencies to the potentia
growth and improvement of this technology.

Now I've referred to the Long Island controversy as a warning sig
nal. It has warned us of several developments in Maine which have seri
ous implications for our environmental quality program.

First, the pressures on our coastal resources are growing by leap
and bounds. We do not have much time in which to plan for their protec
tion.

Secondly, there are a growing number of conflicts, between differen
groups within our cities and towns and between cities and towns, ove
the best use of our land and water resources.

Thirdly, a crisis-to-crisis response to environmental threats is les
and less a satisfactory method for dealing with environmental qualit
problems.

When I was elected Governor in 1954, the major substance of th ssue was economic development. Maine had been plagued by slow growth, fits of isolated prosperity, and regional decline for almost a hundred years. The coastal counties, especially, and particularly those east of Sagadahoc County, were suffering the worst deprivation. And so economic development, just fourteen years ago, became the watchword for every politician, every governor, every candidate for governor, every community across this state. It has been a major goal of every state administration since my first term as governor. And let's confess the truth, that under the urge of that motivation, we've been tempted to sell our resources cheaply—our natural resources and our human resources—for lower wage levels than they're entitled to and our natural resources without any order or care or planning whatsoever.

Until recently, our success in economic development along the coast has been limited. But now proposals are coming up almost too fast for us to react. Consider only these, each of them major: the Long Island development, the nuclear power plant at Wiscasset, the proposed aluminum plant at Trenton and the oil refineries proposals for Machiasport.

Now these are all harbingers of what is to come. They are only the most dramatic signs of change for our state and more of them are coming and they're going to come faster. Are we ready to consider each of them in a concept of some wise public policy related to our total coastal resources? Are we prepared to give yes and no answers to developers who serve us vitally as we think of the opportunity and the public services we want to provide for our children? Are we prepared to make the discriminating judgment which will protect one area against the encroachment of civilization and open another one wisely to the opportunity of industrial growth? These are the kinds of fine, precise judgments that we are developing public policy to meet, because what has started is going to accelerate and build momentum as it has in several hundred miles of coastline to the south of Maine which is overwhelmed with the unhappy consequences.

The inevitable pressures of national population growth, the need for land for residential and industrial expansion, and Maine's unique combination of natural resources are serving as magnets for speculation and development. These great companies are coming here, are seeking to come here, not as a favor to us but because they see opportunities for themselves and more people are going to see it. How are we going to take advantage of the pluses they can bring to us while at the same time minimizing the minuses which their activity has produced?

We see this kind of movement into Maine in industrial proposals, we watch it in land sales as far east as Washington County. Think back fourteen years. Who would have thought then that land in Washington

County would be grabbed by land speculators? For anything! We sens it in the suburban sprawl which is affecting the Portland area and com munities to the north. Each of these developments represents opportu nity, but also poses a threat to the quality of life men and women nee in our state.

The potential abusers of our environment are not just the industria ists, they are the land speculators, home owners, the vacationists, th recreationists, yes, and the hunters and fishermen who count themselve as conservationists. And if you don't believe this, visit the camping site along the storied Allagash.

Many of the arguments over conservation are the products of com peting interests in land and water. Workmen out of jobs or locked in low paying jobs want industrial growth for new opportunity. They're entitle to it. Summer residents and visitors and suburban residents want to limi industrial growth as a protection against interference in their own lif styles. City and town officials try to encourage maximum developmen designed to increase property tax bases. Each of these groups has a differ ent view of what constitutes sound conservation and tends to regard any one in either of the other groups as the natural enemy. Each tends to b willing to sacrifice someone else's interest if he can protect his own.

I think an objective analysis of the Long Island controversy wil reveal conflicts such as this, as will an analysis of any other conservatio fight in Maine in the last few years from the Dickey-Lincoln project t Machiasport. Conservation issues must, of necessity, be decided on case-by-case basis, but they should be decided within the context of ar overall environmental protection policy which takes into account th short-term interest of the various groups of our society and the long-tern interest of society as a whole. They should also take into account the fac that environment is not only the wilderness areas and our rockbound coast. Environment is the community in which we live.

A comprehensive environmental improvement program for Maine must, it seems to me, include at least the following:

1. A substantial reduction in existing air and water pollution. Tha may seem like an obvious statement but you'd be amazed at the extent t which people, including people involved in public policymaking, are will ing to set the floor in other states at the existing level of air and wate pollution. But that isn't good enough and it certainly isn't good enough for Maine, I hope, and it ought to be a deliberate policy, to negate tha kind of an adjustment.

2. A continous effort to improve the quality of our air and water

3. Protection of natural beauty on our coastal and inland waters and their adjoining lands. This may seem simple with 2,500 miles of coastline

f that statistic is sound. I'm not sure that anyone has actually put it to-
ether, but it can disappear so fast.

4. Conservation of our resources against the time of needless ex-
loitation which marred the state through much of the eighteenth, nine-
eenth and early twentieth century.

5. Protection of public interest in the shores and other prime rec-
eational areas of the state.

6. Balanced development of our communities to protect the integrity
f their historic values and to ensure adequate space and public ameni-
ies.

You know, one of the most powerful examples of what can happen
o a community because of its failure to adopt such a balanced develop-
nent policy is our state capital. As I think back to what Augusta looked
ike when I first saw it back in the 1930s, comparing it to what is is today,
can't help but deplore the shortsightedness of leaders, of policymakers,
es and of the city, our capitol city, for not protecting that now long
one, and I'm afraid irrevocably gone, beauty that I associated with the
Augusta of thirty years ago. It had the makings of one of the loveliest
tate capitals in the country but because of the competing interests of
itizens, legitimately pursuing their interests as they saw them, the op-
ortunity to save something beautiful, irreplaceable, has been lost. It
idn't require that Augusta avoid industrialization; it didn't require that
Augusta not provide opportunities for the people or their children. The
wo are consistent.

This is the tragedy of what has happened in so many areas of the
ountry. Beauty and opportunity can go hand in hand. They are con-
istent with each other and, indeed, they enhance each other. And that's
he ultimate truth that we are just beginning to understand.

Maine exists for her people, of course, not for the resources—for her
eople, young and old, rich and poor.

The protection of our resources, for the benefit of her people, de-
nands the foresight and patient devotion exhibited by Governor Baxter
n his life and his legacy.

It will require new institutions for regional planning within the
tate and more explicit authority for state planning, for balanced growth.
Here, again, the Portland area has an opportunity to lead the way. Re-
ional action on a water pollution control program has already begun.
How far it will go without public support may be in doubt. There are
noves in the direction of the regional council of governments to improve
lanning and coordination for the area—essential functions. This is a
ound proposal to institutionalize.

These vehicles can provide the initial opportunity for working out the complicated inter-relationships of land use, tax policies, growth pat terns, and economic needs which effect those decisions we identify a conservation issues. They are complicated. There are no black and white answers to them. We can't lock up our resources in their pristine state i people are to live here and advance; neither can we permit their heedles exploitation. To find the middle ground is the course which demands in terest, concern, self-restraint, understanding of the balanced needs of the community, and understanding of how the other fellow's interests affec our own, and willingness from time to time to sacrifice our own interest or at least to modify them, to take into account the total public good.

I think you have a classic opportunity here to exercise this kind o community citizenship, and community leadership. I urge you, and those who are not here today, not to wait for another Long Island dispute be fore you start work for the protection of your mutual interests.